Template Metaprogramming with C++

Learn everything about C++ templates and unlock the power of template metaprogramming

Marius Bancila

BIRMINGHAM—MUMBAI

Template Metaprogramming with C++

Associate Group Product Manager: Gebin George
Content Development Editor: Rosal Colaco
Technical Editor: Pradeep Sahu
Copy Editor: Safis Editing
Project Coordinator: Manisha Singh
Proofreader: Safis Editing
Indexer: Sejal Dsilva
Production Designer: Vijay Kamble
Business Development Executive: Kriti Sharma
Marketing Coordinator: Sonakshi Bubbar

First published: July 2022

Production reference: 1220722

Published by Packt Publishing Ltd.
Livery Place
35 Livery Street
Birmingham
B3 2PB, UK.

ISBN 978-1-80324-345-0

www.packt.com

To the curious minds that always want to learn more.

– Marius Bancila

Contributors

About the author

Marius Bancila is a software engineer with two decades of experience in developing solutions for line of business applications and more. He is the author of *Modern C++ Programming Cookbook* and *The Modern C++ Challenge*. He works as a software architect and is focused on Microsoft technologies, mainly developing desktop applications with C++ and C#. He is passionate about sharing his technical expertise with others and, for that reason, he has been recognized as a Microsoft MVP for C++ and later developer technologies since 2006. Marius lives in Romania and is active in various online communities.

About the reviewer

Aleksei Goriachikh has 8+ experience in C++ programming. After graduating from Novosibirsk State University in 2012 with a master's degree in mathematics, Aleksei has worked on research projects in the areas of computational mathematics and optimization, a geometric kernel for CAD systems, and a multi-threaded library for autonomous driving. Aleksei's last professional interest is pre-silicon modelling.

Table of Contents

Preface

Part 1: Core Template Concepts

1

An Introduction to Templates

Understanding the need for templates	4	The pros and cons of templates	16
Writing your first templates	8	Summary	17
Understanding template terminology	12	Questions	17
A brief history of templates	14	Further reading	18

2

Template Fundamentals

Defining function templates	20	Understanding template instantiation	43
Defining class templates	23		
Defining member function templates	26	Implicit instantiation	43
		Explicit instantiation	47
Understanding template parameters	28	Understanding template specialization	53
Type template parameters	28	Explicit specialization	53
Non-type template parameters	30	Partial specialization	58
Template template parameters	38		
Default template arguments	40	Defining variable templates	63
		Defining alias templates	67

Exploring generic lambdas
and lambda templates   70
Summary   79

Questions   79
Further reading   79

3

Variadic Templates

Understanding the need
for variadic templates   82
Variadic function templates   84
Parameter packs   90
Understanding parameter
packs expansion   95
Variadic class templates   101

Fold expressions   110
Variadic alias templates   114
Variadic variable templates   116
Summary   117
Questions   117
Further reading   118

Part 2: Advanced Template Features

4

Advanced Template Concepts

Understanding name
binding and dependent names   122
Two-phase name lookup   125
Dependent type names   128
Dependent template names   131
Current instantiation   133

Exploring template recursion 136
Function template
argument deduction   142
Class template
argument deduction   155

Forwarding references   162
The decltype specifier   170
The std::declval type operator 179
Understanding friendship
in templates   181
Summary   188
Questions   189
Further readings   189

5
Type Traits and Conditional Compilation

Understanding and
defining type traits   192
Exploring SFINAE and
its purpose   197
Enabling SFINAE with
the enable_if type trait   203
Using constexpr if   209
Exploring the standard
type traits   213
Querying the type category   214
Querying type properties   217
Querying supported operations   220

Querying type relationships   222
Modifying cv-specifiers, references,
pointers, or a sign   223
Miscellaneous transformations   224

Seeing real-world examples
of using type traits   228
Implementing a copy algorithm   229
Building a homogenous variadic
function template   233
Summary   235
Questions   235
Further reading   235

6
Concepts and Constraints

Understanding the need
for concepts   238
Defining concepts   244
Exploring requires
expressions   248
Simple requirements   248
Type requirements   251
Compound requirements   252
Nested requirements   256

Composing constraints   258
Learning about the ordering
of templates with constraints   262
Constraining non-template
member functions   267

Constraining class templates   271
Constraining variable
templates and
template aliases   273
Learning more ways
to specify constraints   275
Using concepts to constrain
auto parameters   276
Exploring the standard
concepts library   279
Summary   285
Questions   285
Further reading   285

Part 3: Applied Templates

7

Patterns and Idioms

Dynamic versus static polymorphism	290	Alternatives to tag dispatching	332
The Curiously Recurring Template Pattern	294	Expression templates	336
Limiting the object count with CRTP	296	Using ranges as an alternative to expression templates	345
Adding functionality with CRTP	298	Typelists	347
Implementing the composite design pattern	301	Using typelists	349
The CRTP in the standard library	307	Implementing operations on typelists	352
Mixins	312	Summary	361
Type erasure	318	Questions	362
Tag dispatching	328	Further reading	362

8

Ranges and Algorithms

Understanding the design of containers, iterators, and algorithms	366	Writing a custom general-purpose algorithm	395
Creating a custom container and iterator	376	Summary	398
Implementing a circular buffer container	378	Questions	399
Implementing an iterator type for the circular buffer container	387		

9
The Ranges Library

Advancing from abstract ranges to the ranges library	402	Writing your own range adaptor	423
Understanding range concepts and views	404	Summary	432
		Questions	432
Exploring more examples	414	Further reading	433
Understanding the constrained algorithms	419		

Appendix
Closing Notes

Assignment Answers

Chapter 1, Introduction to Templates	437	Chapter 6, Concepts and Constraints	443
Chapter 2, Template Fundamentals	438	Chapter 7, Patterns and Idioms	445
Chapter 3, Variadic Templates	439	Chapter 8, Ranges and Algorithms	446
Chapter 4, Advanced Template Concepts	441	Chapter 9, The Ranges Library	447
Chapter 5, Type Traits and Conditional Compilation	442		

Index

Other Books You May Enjoy

Preface

The C++ programming language is one of the most widely used in the world and it has been so for decades. Its success isn't due just to the performance it provides or maybe to its ease of use, which many would argue against, but probably to its versatility. C++ is a general-purpose, multi-paradigm programming language that blends together procedural, functional, and generic programming.

Generic programming is a paradigm of writing code such as that entities such as functions and classes are written in terms of types that are specified later. These generic entities are instantiated only when needed for specific types that are specified as arguments. These generic entities are known as templates in C++.

Metaprogramming is the programming technique of using templates (and constexpr functions in C++) to generate code at compile-time that is then merged with the rest of the source code for compiling a final program. Metaprogramming implies that at least an input or an output is a type.

Templates in C++ have a reputation of being *pretty horrendous*, as described in the C++ Core Guideless (a document of dos and don'ts maintained by Bjarne Stroustrup and Herb Sutter). However, they make generic libraries possible such as the C++ Standard Library that C++ developers use all the time. Whether you're writing templates yourself or just using templates written by others (such as standard containers or algorithms), templates are most likely part of your daily code.

This book is intended to provide a good understanding of all the spectrum of templates available in C++ (from their basic syntax to concepts in C++20). This will be the focus of the first two parts of the book. The third and final part will help you put the newly acquired knowledge into practice to perform metaprogramming with templates.

Who this book is for?

This book is for beginner-to-intermediate C++ developers who want to learn about template metaprogramming as well as advanced C++ developers looking to get up to speed with the new C++20 features related to templates and the various idioms and patterns. Basic C++ coding experience is necessary to get started with this book.

What this book covers

Chapter 1, Introduction to Templates, provides an introduction to the concept of template metaprogramming in C++, with several simple examples, and a discussion on why we need templates and what are the pros and cons of using templates.

Chapter 2, Template Fundamentals, explores all forms of templates in C++: function templates, class templates, variable templates, and alias templates. For each of these, we discuss the syntax and the details of how they work. Furthermore, the key concepts of template instantiation and specialization are addressed here.

Chapter 3, Variadic Templates, is dedicated entirely to variadic templates which are templates that have a variable number of template parameters. We discuss in detail variadic function templates, variadic class templates, variadic alias templates, and variadic variable templates, parameter packs and how they are expanded, as well as fold expressions that help us simplify the writing of variadic templates.

Chapter 4, Advanced Template Concepts, groups a series of advanced template concepts such as dependent names and name lookup, template argument deduction, template recursion, perfect forwarding, generic and template lambdas. By understanding these topics, readers will be able to greatly expand the variety of templates they can read or write.

Chapter 5, Type Traits and Conditional Compilation, is dedicated to type traits. The reader will learn about type traits, what traits the standard library provides, and how they can be used to solve different problems.

Chapter 6, Concepts and Constraints, presents the new C++20 mechanism for defining requirements for template arguments with concepts and constraints. You will learn about the various ways to specify constraints. Moreover, we provide an overview of the content of the C++20 standard concepts library.

Chapter 7, Patterns and Idioms, explores a series of unrelated advanced topics of using the knowledge learned so far into implementing various patterns. We explore the concepts of static polymorphism, type erasure, tag dispatching, and patterns such as the curiously recursive template pattern, expression templates, mixins, and typelists.

Chapter 8, Ranges and Algorithms, is dedicated to understanding containers, iterators, and algorithms, which are the core components of the standard template library. You will learn here how to write a generic container and an iterator type for it as well as a general-purpose algorithm.

Chapter 9, The Ranges Library, explores the new C++20 Ranges library with its key features such as ranges, range adaptors, and constrained algorithms. These enable us to write simpler code for working with ranges. Furthermore, you will also learn here how to write your own range adaptor.

Appendix is a short epilog that provides a summary of the book.

Assignment Answers contains all the answers to the questions from all the chapters.

To get the most out of this book

To get started with this book, you need to have some basic knowledge of the C++ programming language. You need to know the syntax and fundamentals about classes, functions, operators, function overloading, inheritance, virtual functions, and more. However, no knowledge of templates is required, as this book will teach you everything from scratch.

All the code samples in this book are cross-platform. That means you can use any compiler to build and run them. However, although many snippets work with a C++11 compiler, there are also snippets that require a C++17 or C++20 compliant compiler. Therefore, we recommend you use a compiler version that supports C++20 so you can run all the samples. The samples in this book have been tested with **MSVC 19.30 (Visual Studio 2022)**, **GCC 12.1/13**, and **Clang 13/14**. If you don't have such a C++20 compliant compiler on your machine, you can try one online. We recommend one of the following:

- Compiler Explorer (`https://godbolt.org/`)
- Wandbox (`https://wandbox.org/`)
- C++ Insights (`https://cppinsights.io/`)

The C++ Insights online tools will be referred several times in the book for analyzing the code generated by the compiler.

You should refer to the page, `https://en.cppreference.com/w/cpp/compiler_support`, if you want to check compilers support for different versions of the C++ standard.

If you are using the digital version of this book, we advise you to type the code yourself or access the code from the book's GitHub repository (a link is available in the next section). Doing so will help you avoid any potential errors related to the copying and pasting of code.

Mentions and further readings

Throughout the book, we are referring multiple times to the C++ standard. This document is copy-righted by the **International Organization for Standardization**. The official C++ standard document can be purchased from here: `https://www.iso.org/standard/79358.html`. However, multiple drafts of the C++ standard as well as the sources used to generate them are freely available on GitHub at `https://github.com/cplusplus/draft`. You cand find additional information about the C++ standard at `https://isocpp.org/std/the-standard`.

A great online resource for C++ developers is the **C++ Reference** website, available at `https://en.cppreference.com/`. This provides exhaustive documentation of the C++ language directly derived from the C++ standard. Content from the C++ Reference is quoted several times in the book. The C++ Reference content is licensed under CC-BY-SA license, `https://en.cppreference.com/w/Cppreference:Copyright/CC-BY-SA`.

At the end of each chapter, you will find a section called *Further reading*. This section contains a list of readings used as a bibliography and recommended for deepening your knowledge of the presented topics.

Download the example code files

You can download the example code files for this book from GitHub at `https://github.com/PacktPublishing/Template-Metaprogramming-with-CPP`. If there's an update to the code, it will be updated in the GitHub repository.

We also have other code bundles from our rich catalog of books and videos available at `https://github.com/PacktPublishing/`. Check them out!

Download the color images

We also provide a PDF file that has color images of the screenshots and diagrams used in this book. You can download it here: `https://packt.link/Un8j5`.

Conventions used

There are a number of text conventions used throughout this book.

`Code in text`: Indicates code words in text, database table names, folder names, filenames, file extensions, pathnames, dummy URLs, user input, and Twitter handles. Here is an example: "This problem can be fixed by making `init` a dependent name."

A block of code is set as follows:

```
template <typename T>
struct parser : base_parser<T>
{
    void parse()
    {
        this->init(); // OK
        std::cout << "parse\n";
    }
};
```

Any command-line input or output is written as follows:

```
fatal error: recursive template instantiation exceeded maximum
depth of 1024
use -ftemplate-depth=N to increase recursive template
instantiation depth
```

Bold: Indicates a new term, an important word, or words that you see onscreen. For instance, words in menus or dialog boxes appear in **bold**. Here is an example: "The capacity is 8, the size is 0, and the **head** and **tail** both point to index **0**."

> **Tips or important notes**
> Appear like this.

Get in touch

Feedback from our readers is always welcome.

General feedback: If you have questions about any aspect of this book, email us at customercare@packtpub.com and mention the book title in the subject of your message.

Errata: Although we have taken every care to ensure the accuracy of our content, mistakes do happen. If you have found a mistake in this book, we would be grateful if you would report this to us. Please visit www.packtpub.com/support/errata and fill in the form.

Piracy: If you come across any illegal copies of our works in any form on the internet, we would be grateful if you would provide us with the location address or website name. Please contact us at copyright@packt.com with a link to the material.

If you are interested in becoming an author: If there is a topic that you have expertise in and you are interested in either writing or contributing to a book, please visit authors.packtpub.com.

Share Your Thoughts

Once you've read *Template Metaprogramming with C++*, we'd love to hear your thoughts! Scan the QR code below to go straight to the Amazon review page for this book and share your feedback.

https://packt.link/r/1803243457

Your review is important to us and the tech community and will help us make sure we're delivering excellent quality content.

Part 1: Core Template Concepts

In this part, you will begin with an introduction to templates and understand their benefits. You will then learn about the syntax for writing function templates, class templates, variable templates, and alias templates. You will explore concepts such as template instantiation and template specialization and learn how to write templates with a variable number of arguments.

This part comprises the following chapters:

- *Chapter 1, Introduction to Templates*
- *Chapter 2, Template Fundamentals*
- *Chapter 3, Variadic Templates*

1

An Introduction to Templates

As a C++ developer, you should be at least familiar if not well versed in **template metaprogramming**, in general, referred to in short as **templates**. Template metaprogramming is a programming technique that uses templates as blueprints for the compiler to generate code and help developers avoid writing repetitive code. Although general-purpose libraries use templates heavily, the syntax and the inner workings of templates in the C++ language can be discouraging. Even C++ *Core Guidelines*, which is a collection of dos and don'ts edited by *Bjarne Stroustrup*, the creator of the C++ language, and *Herb Sutter*, the chair of the C++ standardization committee, calls templates *pretty horrendous*.

This book is intended to shed light on this area of the C++ language and help you become prolific in template metaprogramming.

In this chapter, we will go through the following topics:

- Understanding the need for templates
- Writing your first templates
- Understanding template terminology
- A brief history of templates
- The pros and cons of templates

The first step in learning how to use templates is to understand what problem they actually solve. Let's start with that.

Understanding the need for templates

Each language feature is designed to help with a problem or task that developers face when using that language. The purpose of templates is to help us avoid writing repetitive code that only differs slightly.

To exemplify this, let's take the classical example of a max function. Such a function takes two numerical arguments and returns the largest of the two. We can easily implement this as follows:

```
int max(int const a, int const b)
{
    return a > b ? a : b;
}
```

This works pretty well, but as you can see, it will only work for values of the type int (or those that are convertible to int). What if we need the same function but with arguments of the type double? Then, we can overload this function (create a function with the same name but a different number or type of arguments) for the double type:

```
double max(double const a, double const b)
{
    return a > b ? a : b;
}
```

However, int and double are not the only numeric types. There is char, short, long, long and their unsigned counterparts, unsigned char, unsigned short, unsigned long, and unsigned long. There are also the types float and long double. And other types, such as int8_t, int16_t, int32_t, and int64_t. And there could be other types that can be compared, such as bigint, Matrix, point2d, and any user-defined type that overloads operator>. How can a general-purpose library provide a general-purpose function such as max for all these types? It can overload the function for all the built-in types and perhaps other library types but cannot do so for any user-defined type.

An alternative to overloading functions with different parameters is to use `void*` to pass arguments of different types. Keep in mind this is a bad practice and the following example is shown only as a possible alternative in a world without templates. However, for the sake of discussion, we can design a sorting function that will run the quick sort algorithm on an array of elements of any possible type that provides a strict weak ordering. The details of the quicksort algorithm can be looked up online, such as on Wikipedia at `https://en.wikipedia.org/wiki/Quicksort`.

The quicksort algorithm needs to compare and swap any two elements. However, since we don't know their type, the implementation cannot do this directly. The solution is to rely on **callbacks**, which are functions passed as arguments that will be invoked when necessary. A possible implementation can be as follows:

```
using swap_fn = void(*)(void*, int const, int const);
using compare_fn = bool(*)(void*, int const, int const);

int partition(void* arr, int const low, int const high,
              compare_fn fcomp, swap_fn fswap)
{
    int i = low - 1;

    for (int j = low; j <= high - 1; j++)
    {
        if (fcomp(arr, j, high))
        {
            i++;
            fswap(arr, i, j);
        }
    }

    fswap(arr, i + 1, high);

    return i + 1;
}

void quicksort(void* arr, int const low, int const high,
               compare_fn fcomp, swap_fn fswap)
{
```

```
    if (low < high)
    {
        int const pi = partition(arr, low, high, fcomp,
            fswap);
        quicksort(arr, low, pi - 1, fcomp, fswap);
        quicksort(arr, pi + 1, high, fcomp, fswap);
    }
}
```

In order to invoke the `quicksort` function, we need to provide implementations for these comparisons and swapping functions for each type of array that we pass to the function. The following are implementations for the `int` type:

```
void swap_int(void* arr, int const i, int const j)
{
    int* iarr = (int*)arr;
    int t = iarr[i];
    iarr[i] = iarr[j];
    iarr[j] = t;
}

bool less_int(void* arr, int const i, int const j)
{
    int* iarr = (int*)arr;
    return iarr[i] <= iarr[j];
}
```

With all these defined, we can write code that sorts arrays of integers as follows:

```
int main()
{
    int arr[] = { 13, 1, 8, 3, 5, 2, 1 };
    int n = sizeof(arr) / sizeof(arr[0]);
    quicksort(arr, 0, n - 1, less_int, swap_int);
}
```

These examples focused on functions but the same problem applies to classes. Consider that you want to write a class that models a collection of numerical values that has variable size and stores the elements contiguously in memory. You could provide the following implementation (only the declaration is sketched here) for storing integers:

```
struct int_vector
{
    int_vector();

    size_t size() const;
    size_t capacity() const;
    bool empty() const;

    void clear();
    void resize(size_t const size);

    void push_back(int value);
    void pop_back();

    int at(size_t const index) const;
    int operator[](size_t const index) const;
private:
    int* data_;
    size_t size_;
    size_t capacity_;
};
```

This all looks good but the moment you need to store values of the type `double`, or `std::string`, or any user-defined type you'll have to write the same code, each time only changing the type of the elements. This is something nobody wants to do because it is repetitive work and because when something needs to change (such as adding a new feature or fixing a bug) you need to apply the same change in multiple places.

Lastly, a similar problem can be encountered, although less often, when you need to define variables. Let's consider the case of a variable that holds the new line character. You can declare it as follows:

```
constexpr char NewLine = '\n';
```

What if you need the same constant but for a different encoding, such as wide string literals, UTF-8, and so on? You can have multiple variables, having different names, such as in the following example:

```
constexpr wchar_t NewLineW = L'\n';
constexpr char8_t NewLineU8 = u8'\n';
constexpr char16_t NewLineU16 = u'\n';
constexpr char32_t NewLineU32 = U'\n';
```

Templates are a technique that allows developers to write blueprints that enable the compiler to generate all this repetitive code for us. In the following section, we will see how to transform the preceding snippets into C++ templates.

Writing your first templates

It is now time to see how templates are written in the C++ language. In this section, we will start with three simple examples, one for each of the snippets presented earlier.

A template version of the max function discussed previously would look as follows:

```
template <typename T>
T max(T const a, T const b)
{
    return a > b ? a : b;
}
```

You will notice here that the type name (such as int or double) has been replaced with T (which stands for *type*). T is called a **type template parameter** and is introduced with the syntax template<typename T> or typename<class T>. Keep in mind that T is a parameter, therefore it can have any name. We will learn more about template parameters in the next chapter.

At this point, this template that you put in the source code is only a blueprint. The compiler will generate code from it based on its use. More precisely, it will instantiate a function overload for each type the template is used with. Here is an example:

```
struct foo{};

int main()
{
    foo f1, f2;
```

```
    max(1, 2);       // OK, compares ints
    max(1.0, 2.0);   // OK, compares doubles
    max(f1, f2);     // Error, operator> not overloaded for
                     // foo
}
```

In this snippet, we are first calling max with two integers, which is OK because operator> is available for the type int. This will generate an overload int max(int const a, int const b). Second, we are calling max with two doubles, which again is all right since operator> works for doubles. Therefore, the compiler will generate another overload, double max(double const a, double const b). However, the third call to max will generate a compiler error, because the foo type does not have the operator> overloaded.

Without getting into too many details at this point, it should be mentioned that the complete syntax for calling the max function is the following:

```
max<int>(1, 2);
max<double>(1.0, 2.0);
max<foo>(f1, f2);
```

The compiler is able to deduce the type of the template parameter, making it redundant to write it. There are cases, however, when that is not possible; in those situations, you need to specify the type explicitly, using this syntax.

The second example involving functions from the previous section, *Understanding the need for templates*, was the quicksort() implementation that dealt with void* arguments. The implementation can be easily transformed into a template version with very few changes. This is shown in the following snippet:

```
template <typename T>
void swap(T* a, T* b)
{
    T t = *a;
    *a = *b;
    *b = t;
}


template <typename T>
int partition(T arr[], int const low, int const high)
```

```
{
    T pivot = arr[high];
    int i = (low - 1);


    for (int j = low; j <= high - 1; j++)
    {
        if (arr[j] < pivot)
        {
            i++;
            swap(&arr[i], &arr[j]);
        }
    }


    swap(&arr[i + 1], &arr[high]);


    return i + 1;
}


template <typename T>
void quicksort(T arr[], int const low, int const high)
{
    if (low < high)
    {
        int const pi = partition(arr, low, high);
        quicksort(arr, low, pi - 1);
        quicksort(arr, pi + 1, high);
    }
}
```

The use of the quicksort function template is very similar to what we have seen earlier, except there is no need to pass pointers to callback functions:

```
int main()
{
    int arr[] = { 13, 1, 8, 3, 5, 2, 1 };
```

```
    int n = sizeof(arr) / sizeof(arr[0]);
    quicksort(arr, 0, n - 1);
}
```

The third example we looked at in the previous section was the `vector` class. A template version of it will look as follows:

```
template <typename T>
struct vector
{
    vector();

    size_t size() const;
    size_t capacity() const;
    bool empty() const;

    void clear();
    void resize(size_t const size);

    void push_back(T value);
    void pop_back();

    T at(size_t const index) const;
    T operator[](size_t const index) const;
private:
    T* data_;
    size_t size_;
    size_t capacity_;
};
```

As in the case of the `max` function, the changes are minimal. There is the template declaration on the line above the class and the type `int` of the elements has been replaced with the type template parameter T. This implementation can be used as follows:

```
int main()
{
    vector<int> v;
```

```
    v.push_back(1);
    v.push_back(2);
}
```

One thing to notice here is that we have to specify the type of the elements when declaring the variable v, which is int in our snippet because the compiler would not be able to infer their type otherwise. There are cases when this is possible, in C++17, and this topic, called **class template argument deduction**, will be discussed in *Chapter 4, Advanced Template Concepts*.

The fourth and last example concerned the declaration of several variables when only the type was different. We could replace all those variables with a template, as shown in the following snippet:

```
template<typename T>
constexpr T NewLine = T('\n');
```

This template can be used as follows:

```
int main()
{
    std::wstring test = L"demo";
    test += NewLine<wchar_t>;
    std::wcout << test;
}
```

The examples in this section show that the syntax for declaring and using templates is the same whether they represent functions, classes, or variables. This leads us to the next section where we will discuss the types of templates and template terminology.

Understanding template terminology

So far in this chapter, we have used the general term templates. However, there are four different terms describing the kind of templates we have written:

- **Function template** is the term used for a templated function. An example is the max template seen previously.

- **Class template** is the term used for a templated class (which can be defined either with the class, struct, or union keyword). An example is the vector class we wrote in the previous section.

- **Variable template** is the term used for templated variables, such as the `NewLine` template from the previous section.

- **Alias template** is the term used for templated type aliases. We will see examples for alias templates in the next chapter.

Templates are parameterized with one or more parameters (in the examples we have seen so far, there was a single parameter). These are called **template parameters** and can be of three categories:

- **Type template parameters**, such as in `template<typename T>`, where the parameter represents a type specified when the template is used.

- **Non-type template parameters**, such as in `template<size_t N>` or `template<auto n>`, where each parameter must have a structural type, which includes integral types, floating-point types (as for C++20), pointer types, enumeration types, lvalue reference types, and others.

- **Template template parameters**, such as in `template<typename K, typename V, template<typename> typename C>`, where the type of a parameter is another template.

Templates can be specialized by providing alternative implementations. These implementations can depend on the characteristics of the template parameters. The purpose of specialization is to enable optimizations or reduce code bloat. There are two forms of specialization:

- **Partial specialization**: This is an alternative implementation provided for only some of the template parameters.

- **(Explicit) full specialization**: This is a specialization of a template when all the template arguments are provided.

The process of generating code from a template by the compiler is called **template instantiation**. This happens by substituting the template arguments for the template parameters used in the definition of the template. For instance, in the example where we used `vector<int>`, the compiler substituted the `int` type in every place where `T` appeared.

Template instantiation can have two forms:

- **Implicit instantiation**: This occurs when the compiler instantiates a template due to its use in code. This happens only for those combinations or arguments that are in use. For instance, if the compiler encounters the use of `vector<int>` and `vector<double>`, it will instantiate the `vector` class template for the types `int` and `double` and nothing more.

- **Explicit instantiation**: This is a way to explicitly tell the compiler what instantiations of a template to create, even if those instantiations are not explicitly used in your code. This is useful, for instance, when creating library files, because uninstantiated templates are not put into object files. They also help reduce compile times and object sizes, in ways that we will see at a later time.

All the terms and topics mentioned in this section will be detailed in other chapters of the book. This section is intended as a short reference guide to template terminology. Keep in mind though that there are many other terms related to templates that will be introduced at the appropriate time.

A brief history of templates

Template metaprogramming is the C++ implementation of generic programming. This paradigm was first explored in the 1970s and the first major languages to support it were Ada and Eiffel in the first half of the 1980s. David Musser and Alexander Stepanov defined generic programming, in a paper called *Generic Programming*, in 1989, as follows:

> *Generic programming centers around the idea of abstracting from concrete, efficient algorithms to obtain generic algorithms that can be combined with different data representations to produce a wide variety of useful software.*

This defines a paradigm of programming where algorithms are defined in terms of types that are specified later and instantiated based on their use.

Templates were not part of the initial **C with Classes** language developed by Bjarne Stroustrup. Stroustrup's first papers describing templates in C++ appeared in 1986, one year after the publication of his book, *The C++ Programming Language, First Edition*. Templates became available in the C++ language in 1990, before the ANSI and ISO C++ committees for standardization were founded.

In the early 1990s, Alexander Stepanov, David Musser, and Meng Lee experimented with the implementation in C++ of various generic concepts. This led to the first implementation of the **Standard Template Library** (**STL**). When the ANSI/ISO committee became aware of the library in 1994, it quickly added it to the drafted specifications. STL was standardized along with the C++ language in 1998 in what became known as C++98.

Newer versions of the C++ standard, collectively referred to as **modern C++**, have introduced various improvements to template metaprogramming. The following table lists them briefly:

Version	Feature	Description
C++11	Variadic templates	Templates can have a variable number of template parameters.
	Template aliases	Ability to define synonyms for a template type with the help of using declarations.
	Extern templates	To tell the compiler not to instantiate a template in a translation unit.
	Type traits	The new header `<type_traits>` contains standard type traits that identify the category of an object and characteristics of a type.
C++14	Variable templates	Support for defining variables or static data members that are templates.
C++17	Fold expressions	Reduce the parameter pack of a variadic template over a binary operator.
	`typename` in template parameters	The `typename` keyword can be used instead of class in a template parameter.
	`auto` for non-type template parameter	The keyword `auto` can be used for non-type template parameters.
	Class template argument deduction	The compiler infers the type of template parameters from the way an object is initialized.
C++20	Template lambdas	Lambdas can be templates just like regular functions.
	String literals as template parameters	String literals can be used as non-type template arguments and a new form of the user-defined literal operator for strings.
	Constraints	Define requirements on template arguments.
	Concepts	Named sets of constraints.

Table 1.1

All these features, along with other aspects of template metaprogramming, will make the sole subject of this book and will be presented in detail in the following chapters. For now, let's see what the advantages and disadvantages are of using templates.

The pros and cons of templates

Before you start using templates, it's important to understand the benefits of using them as well as the disadvantages they may incur.

Let's start by pointing out the advantages:

- Templates help us avoid writing repetitive code.

- Templates foster the creation of generic libraries providing algorithms and types, such as the standard C++ library (sometimes incorrectly referred to as the STL), which can be used in many applications, regardless of their type.

- The use of templates can result in less and better code. For instance, using algorithms from the standard library can help write less code that is likely easier to understand and maintain and also probably more robust because of the effort put into the development and testing of these algorithms.

When it comes to disadvantages, the following are worth mentioning:

- The syntax is considered complex and cumbersome, although with a little practice this should not really pose a real hurdle in the development and use of templates.

- Compiler errors related to template code can often be long and cryptic, making it very hard to identify their cause. Newer versions of the C++ compilers have made progress in simplifying these kinds of errors, although they generally remain an important issue. The inclusion of concepts in the C++20 standard has been seen as an attempt, among others, to help provide better diagnostics for compiling errors.

- They increase the compilation times because they are implemented entirely in headers. Whenever a change to a template is made, all the translation units in which that header is included must be recompiled.

- Template libraries are provided as a collection of one or more headers that must be compiled together with the code that uses them.

- Another disadvantage that results from the implementation of templates in headers is that there is no information hiding. The entire template code is available in headers for anyone to read. Library developers often resort to the use of namespaces with names such as `detail` or `details` to contain code that is supposed to be internal for a library and should not be called directly by those using the library.

- They could be harder to validate since code that is not used is not instantiated by the compiler. It is, therefore, important that when writing unit tests, good code coverage must be ensured. This is especially the case for libraries.

Although the list of disadvantages may seem longer, the use of templates is not a bad thing or something to be avoided. On the contrary, templates are a powerful feature of the C++ language. Templates are not always properly understood and sometimes are misused or overused. However, the judicious use of templates has unquestionable advantages. This book will try to provide a better understanding of templates and their use.

Summary

This chapter introduced the concept of templates in the C++ programming language.

We started by learning about the problems for which the solution is the use of templates. We then saw how templates look with simple examples of function templates, class templates, and variable templates. We introduced the basic terminology for templates, which we will discuss more in the forthcoming chapters. Toward the end of the chapter, we saw a brief history of templates in the C++ programming language. We ended the chapter with a discussion on the advantages and disadvantages of using templates. All these topics will lead us to understand the next chapters better.

In the next chapter, we will explore the fundamentals of templates in C++.

Questions

1. Why do we need templates? What advantages do they provide?
2. How do you call a function that is a template? What about a class that is a template?
3. How many kinds of template parameters exist and what are they?
4. What is partial specialization? What about full specialization?
5. What are the main disadvantages of using templates?

Further reading

- *Generic Programming, David Musser, Alexander Stepanov,* `http://stepanovpapers.com/genprog.pdf`

- *A History of C++: 1979–1991, Bjarne Stroustrup,* `https://www.stroustrup.com/hopl2.pdf`

- *History of C++,* `https://en.cppreference.com/w/cpp/language/history`

- *Templates in C++ - Pros and Cons, Sergey Chepurin,* `https://www.codeproject.com/Articles/275063/Templates-in-Cplusplus-Pros-and-Cons`

2
Template Fundamentals

In the previous chapter, we saw a short introduction to templates. What they are, how they are helpful, pros and cons for using templates, and, also, a few examples of function and class templates. In this chapter, we will explore this area in detail, and look at aspects such as template parameters, instantiation, specializations, aliases, and more. The main topics that you will learn about from this chapter are as follows:

- How to define function templates, class templates, variable templates, and alias templates

- What kinds of template parameters exist?

- What is template instantiation?

- What is template specialization?

- How to use generic lambdas and lambda templates

By the end of this chapter, you will be familiar with the core fundamentals of templates in C++ and be able to understand large areas of template code and also write templates by yourself.

To start this chapter, we will explore the details of defining and using function templates.

Defining function templates

Function templates are defined in a similar way to regular functions, except that the function declaration is preceded by the keyword `template` followed by a list of template parameters between angle brackets. The following is a simple example of a function template:

```
template <typename T>
T add(T const a, T const b)
{
    return a + b;
}
```

This function has two parameters, called a and b, both of the same T type. This type is listed in the template parameters list, introduced with the keyword `typename` or `class` (the former is used in this example and throughout the book). This function does nothing more than add the two arguments and returns the result of this operation, which should have the same T type.

Function templates are only blueprints for creating actual functions and only exist in source code. Unless explicitly called in your source code, the function templates will not be present in the compiled executable. However, when the compiler encounters a call to a function template and is able to match the supplied arguments and their types to a function template's parameters, it generates an actual function from the template and the arguments used to invoke it. To understand this, let's look at some examples:

```
auto a = add(42, 21);
```

In this snippet, we call the add function with two `int` parameters, 42 and 21. The compiler is able to deduce the template parameter T from the type of the supplied arguments, making it unnecessary to explicitly provide it. However, the following two invocations are also possible, and, in fact, identical to the earlier one:

```
auto a = add<int>(42, 21);
auto a = add<>(42, 21);
```

From this invocation, the compiler will generate the following function (keep in mind that the actual code may differ for various compilers):

```
int add(const int a, const int b)
{
    return a + b;
}
```

However, if we change the call to the following form, we explicitly provide the argument for the template parameter T, as the short type:

```
auto b = add<short>(42, 21);
```

In this case, the compiler will generate another instantiation of this function, with short instead of int. This new instantiation would look as follows:

```
short add(const short a, const int b)
{
    return static_cast<short>(a + b);
}
```

If the type of the two parameters is ambiguous, the compiler will not be able to deduce them automatically. This is the case with the following invocation:

```
auto d = add(41.0, 21);
```

In this example, 41.0 is a double but 21 is an int. The add function template has two parameters of the same type, so the compiler is not able to match it with the supplied arguments and will issue an error. To avoid this, and suppose you expected it to be instantiated for double, you have to specify the type explicitly, as shown in the following snippet:

```
auto d = add<double>(41.0, 21);
```

As long as the two arguments have the same type and the + operator is available for the type of the arguments, you can call the function template add in the ways shown previously. However, if the + operator is not available, then the compiler will not be able to generate an instantiation, even if the template parameters are correctly resolved. This is shown in the following snippet:

```
class foo
{
    int value;
public:
    explicit foo(int const i):value(i)
    { }

    explicit operator int() const { return value; }
```

```
};
```

```
auto f = add(foo(42), foo(41));
```

In this case, the compiler will issue an error that a binary + operator is not found for arguments of type foo. Of course, the actual message differs for different compilers, which is the case for all errors. To make it possible to call add for arguments of type foo, you'd have to overload the + operator for this type. A possible implementation is the following:

```
foo operator+(foo const a, foo const b)
{
   return foo((int)a + (int)b);
}
```

All the examples that we have seen so far represented templates with a single template parameter. However, a template can have any number of parameters and even a variable number of parameters. This latter topic will be addressed in *Chapter 3, Variadic Templates*. The next function is a function template that has two type template parameters:

```
template <typename Input, typename Predicate>
int count_if(Input start, Input end, Predicate p)
{
   int total = 0;
   for (Input i = start; i != end; i++)
   {
      if (p(*i))
         total++;
   }
   return total;
}
```

This function takes two input iterators to the start and end of a range and a predicate and returns the number of elements in the range that match the predicate. This function, at least conceptually, is very similar to the std::count_if general-purpose function from the <algorithm> header in the standard library and you should always prefer to use standard algorithms over hand-crafted implementations. However, for the purpose of this topic, this function is a good example to help you understand how templates work.

We can use the `count_if` function as follows:

```
int main()
{
    int arr[]{ 1,1,2,3,5,8,11 };
    int odds = count_if(
                    std::begin(arr), std::end(arr),
                    [](int const n) { return n % 2 == 1; });
    std::cout << odds << '\n';
}
```

Again, there is no need to explicitly specify the arguments for the type template parameters (the type of the input iterator and the type of the unary predicate) because the compiler is able to infer them from the call.

Although there are more things to learn about function templates, this section provided an introduction to working with them. Let's now learn the basics of defining class templates.

Defining class templates

Class templates are declared in a very similar manner, with the `template` keyword and the template parameter list preceding the class declaration. We saw the first example in the introductory chapter. The next snippet shows a class template called `wrapper`. It has a single template parameter, a type called `T`, that is used as the type for data members, parameters, and function return types:

```
template <typename T>
class wrapper
{
public:
    wrapper(T const v) : value(v)
    { }

    T const& get() const { return value; }
private:
    T value;
};
```

As long as the class template is not used anywhere in your source code, the compiler will not generate code from it. For that to happen, the class template must be instantiated and all its parameters properly matched to arguments either explicitly, by the user, or implicitly, by the compiler. Examples for instantiating this class template are shown next:

```
wrapper a(42);              // wraps an int
wrapper<int> b(42);         // wraps an int
wrapper<short> c(42);       // wraps a short
wrapper<double> d(42.0);    // wraps a double
wrapper e("42");            // wraps a char const *
```

The definitions of a and e in this snippet are only valid in C++17 and onward thanks to a feature called **class template argument deduction**. This feature enables us to use class templates without specifying any template argument, as long as the compiler is able to deduce them all. This will be discussed in *Chapter 4, Advanced Template Concepts*. Until then, all examples that refer to class templates will explicitly list the arguments, as in wrapper<int> or wrapper<char const*>.

Class templates can be declared without being defined and used in contexts where incomplete types are allowed, such as the declaration of a function, as shown here:

```
template <typename T>
class wrapper;


void use_foo(wrapper<int>* ptr);
```

However, a class template must be defined at the point where the template instantiation occurs; otherwise, the compiler will generate an error. This is exemplified with the following snippet:

```
template <typename T>
class wrapper;                                // OK


void use_wrapper(wrapper<int>* ptr); // OK


int main()
{
    wrapper<int> a(42);                       // error, incomplete type
    use_wrapper(&a);
```

```
}


template <typename T>
class wrapper
{
    // template definition
};


void use_wrapper(wrapper<int>* ptr)
{
    std::cout << ptr->get() << '\n';
}
```

When declaring the use_wrapper function, the class template wrapper is only declared, but not defined. However, incomplete types are allowed in this context, which makes it all right to use wrapper<T> at this point. However, in the main function we are instantiating an object of the wrapper class template. This will generate a compiler error because at this point the definition of the class template must be available. To fix this particular example, we'd have to move the definition of the main function to the end, after the definition of wrapper and use_wrapper.

In this example, the class template was defined using the class keyword. However, in C++ there is little difference between declaring classes with the class or struct keyword:

- With struct, the default member access is public, whereas using class is private.
- With struct, the default access specifier for base-class inheritance is public, whereas using class is private.

You can define class templates using the struct keyword the same way we did here using the class keyword. The differences between classes defined with the struct or the class keyword are also observed for class templates defined with the struct or class keyword.

Classes, whether they are templates or not, may contain member function templates too. The way these are defined is discussed in the next section.

Defining member function templates

So far, we have learned about function templates and class templates. It is possible to define member function templates too, both in non-template classes and class templates. In this section, we will learn how to do this. To understand the differences, let's start with the following example:

```
template <typename T>
class composition
{
public:
    T add(T const a, T const b)
    {
        return a + b;
    }
};
```

The composition class is a class template. It has a single member function called add that uses the type parameter T. This class can be used as follows:

```
composition<int> c;
c.add(41, 21);
```

We first need to instantiate an object of the composition class. Notice that we must explicitly specify the argument for the type parameter T because the compiler is not able to figure it out by itself (there is no context from which to infer it). When we invoke the add function, we just provide the arguments. Their type, represented by the T type template parameter that was previously resolved to int, is already known. A call such as c.add<int>(42, 21) would trigger a compiler error. The add function is not a function template, but a regular function that is a member of the composition class template.

In the next example, the composition class changes slightly, but significantly. Let's see the definition first:

```
class composition
{
public:
    template <typename T>
    T add(T const a, T const b)
    {
```

```
        return a + b;
    }
};
```

This time, `composition` is a non-template class. However, the `add` function is a function template. Therefore, to call this function, we must do the following:

```
composition c;
c.add<int>(41, 21);
```

The explicit specification of the `int` type for the `T` type template parameter is redundant since the compiler can deduce it by itself from the arguments of the call. However, it was shown here to better comprehend the differences between these two implementations.

Apart from these two cases, member functions of class templates and member function templates of classes, we can also have member function templates of class templates. In this case, however, the template parameters of the member function template must differ from the template parameters of the class template; otherwise, the compiler will generate an error. Let's return to the `wrapper` class template example and modify it as follows:

```
template <typename T>
class wrapper
{
public:
    wrapper(T const v) :value(v)
    {}

    T const& get() const { return value; }

    template <typename U>
    U as() const
    {
        return static_cast<U>(value);
    }
private:
    T value;
};
```

As you can see here, this implementation features one more member, a function called `as`. This is a function template and has a type template parameter called `U`. This function is used to cast the wrapped value from a type `T` to a type `U`, and return it to the caller. We can use this implementation as follows:

```
wrapper<double> a(42.0);
auto d = a.get();         // double
auto n = a.as<int>();     // int
```

Arguments for the template parameters were specified when instantiating the `wrapper` class (`double`) – although in C++17 this is redundant, and when invoking the `as` function (`int`) to perform the cast.

Before we continue with other topics such as instantiation, specialization, and other forms of templates, including variables and aliases, it's important that we take the time to learn more about template parameters. This will make the subject of the next section.

Understanding template parameters

So far in the book, we have seen multiple examples of templates with one or more parameters. In all these examples, the parameters represented types supplied at instantiation, either explicitly by the user, or implicitly by the compiler when it could deduce them. These kinds of parameters are called **type template parameters**. However, templates can also have **non-type template parameters** and **template template parameters**. In the following sections, we'll explore all of them.

Type template parameters

As already mentioned, these are parameters representing types supplied as arguments during the template instantiation. They are introduced either with the `typename` or the `class` keyword. There is no difference between using these two keywords. A type template parameter can have a default value, which is a type. This is specified the same way you would specify a default value for a function parameter. Examples for these are shown in the following snippet:

```
template <typename T>
class wrapper { /* ... */ };


template <typename T = int>
class wrapper { /* ... */ };
```

The name of the type template parameter can be omitted, which can be useful in forwarding declarations:

```
template <typename>
class wrapper;


template <typename = int>
class wrapper;
```

C++11 has introduced variadic templates, which are templates with a variable number of arguments. A template parameter that accepts zero or more arguments is called a **parameter pack**. A **type template parameter pack** has the following form:

```
template <typename... T>
class wrapper { /* ... */ };
```

Variadic templates will be addressed in *Chapter 3, Variadic Templates*. Therefore, we will not get into details about these kinds of parameters at this point.

C++20 has introduced **concepts** and **constraints**. Constraints specify requirements on template arguments. A named set of constraints is called a concept. Concepts can be specified as type template parameters. However, the syntax is a little bit different. The name of the concept (followed by a list of template arguments in angle brackets if the case) is used instead of the typename or the class keyword. Examples, including concepts with a default value and constrained type template parameter pack, are shown as follows:

```
template <WrappableType T>
class wrapper { /* ... */ };


template <WrappableType T = int>
class wrapper { /* ... */ };


template <WrappableType... T>
class wrapper { /* ... */ };
```

Concepts and constraints are discussed in *Chapter 6, Concepts and Constraints*. We will learn more about these kinds of parameters in that chapter. For now, let's look at the second kind of template parameters, non-type template parameters.

Non-type template parameters

Template arguments don't always have to represent types. They can also be compile-time expressions, such as constants, addresses of functions or objects with external linkage, or addresses of static class members. Parameters supplied with compile-time expressions are called **non-type template parameters**. This category of parameters can only have a **structural type**. The following are the structural types:

- Integral types

- Floating-point types, as of C++20

- Enumerations

- Pointer types (either to objects or functions)

- Pointer to member types (either to member objects or member functions)

- Lvalue reference types (either to objects or functions)

- A literal class type that meets the following requirements:

  - All base classes are public and non-mutable.

  - All non-static data members are public and non-mutable.

  - The types of all base classes and non-static data members are also structural types or arrays thereof.

cv-qualified forms of these types can also be used for non-type template parameters. Non-type template parameters can be specified in different ways. The possible forms are shown in the following snippet:

```
template <int V>
class foo { /*...*/ };


template <int V = 42>
class foo { /*...*/ };


template <int... V>
class foo { /*...*/ };
```

In all these examples, the type of the non-type template parameters is `int`. The first and second examples are similar, except that in the second example a default value is used. The third example is significantly different because the parameter is actually a parameter pack. This will be discussed in the next chapter.

To understand non-type template parameters better, let's look at the following example, where we sketch a fixed-size array class, called `buffer`:

```
template <typename T, size_t S>
class buffer
{
    T data_[S];
public:
    constexpr T const * data() const { return data_; }

    constexpr T& operator[](size_t const index)
    {
        return data_[index];
    }

    constexpr T const & operator[](size_t const index) const
    {
        return data_[index];
    }
};
```

This `buffer` class holds an internal array of S elements of type T. Therefore, S needs to be a compile-type value. This class can be instantiated as follows:

```
buffer<int, 10> b1;
buffer<int, 2*5> b2;
```

These two definitions are equivalent, and both b1 and b2 are two buffers holding 10 integers. Moreover, they are of the same type, since 2*5 and 10 are two expressions evaluated to the same compile-time value. You can easily check this with the following statement:

```
static_assert(std::is_same_v<decltype(b1), decltype(b2)>);
```

This is not the case anymore, for the type of the b3 object is declared as follows:

```
buffer<int, 3*5> b3;
```

In this example, b3 is a buffer holding 15 integers, which is different from the buffer type from the previous example that held 10 integers. Conceptually, the compiler generates the following code:

```
template <typename T, size_t S>
class buffer
{
    T data_[S];
public:
    constexpr T* data() const { return data_; }

    constexpr T& operator[](size_t const index)
    {
        return data_[index];
    }

    constexpr T const & operator[](size_t const index) const
    {
        return data_[index];
    }
};
```

This is the code for the primary template but there are also a couple of specializations shown next:

```
template<>
class buffer<int, 10>
{
    int data_[10];
public:
    constexpr int * data() const;
    constexpr int & operator[](const size_t index);
    constexpr const int & operator[](
        const size_t index) const;
};

template<>
class buffer<int, 15>
```

```
{
   int data_[15];
public:
   constexpr int * data() const;
   constexpr int & operator[](const size_t index);
   constexpr const int & operator[](
      const size_t index) const;
};
```

The concept of specialization, seen in this code sample, is detailed further on in this chapter, in the *Understanding template specialization* section. For the time being, you should notice the two different buffer types. Again, it's possible to verify that the types of b1 and b3 are different with the following statement:

```
static_assert(!std::is_same_v<decltype(b1), decltype(b3)>);
```

The use of structural types such as integer, floating-point, or enumeration types is encountered in practice more often than the rest. It's probably easier to understand their use and find useful examples for them. However, there are scenarios where pointers or references are used. In the following example, we will examine the use of a pointer to function parameter. Let's see the code first:

```
struct device
{
    virtual void output() = 0;
    virtual ~device() {}
};

template <void (*action)()>
struct smart_device : device
{
    void output() override
    {
        (*action)();
    }
};
```

In this snippet, `device` is a base class with a pure virtual function called `output` (and a virtual destructor). This is the base class for a class template called `smart_device` that implements the `output` virtual function by calling a function through a function pointer. This function pointer is passed an argument for the non-type template parameter of the class template. The following sample shows how it can be used:

```
void say_hello_in_english()
{
    std::cout << "Hello, world!\n";
}

void say_hello_in_spanish()
{
    std::cout << "Hola mundo!\n";
}

auto w1 =
    std::make_unique<smart_device<&say_hello_in_english>>();
w1->output();

auto w2 =
    std::make_unique<smart_device<&say_hello_in_spanish>>();
w2->output();
```

Here, w1 and w2 are two `unique_ptr` objects. Although, apparently, they point to objects of the same type, that is not true, because `smart_device<&say_hello_in_english>` and `smart_device<&say_hello_in_spanish>` are different types since they are instantiated with different values for the function pointer. This can be easily checked with the following statement:

```
static_assert(!std::is_same_v<decltype(w1), decltype(w2)>);
```

If we, on the other hand, change the `auto` specifier with `std::unique_ptr<device>`, as shown in the following snippet, then w1 and w2 are smart pointers to the base class device, and therefore have the same type:

```
std::unique_ptr<device> w1 =
    std::make_unique<smart_device<&say_hello_in_english>>();
w1->output();
```

```
std::unique_ptr<device> w2 =
    std::make_unique<smart_device<&say_hello_in_spanish>>();
w2->output();


static_assert(std::is_same_v<decltype(w1), decltype(w2)>);
```

Although this example uses a pointer to function, a similar example can be conceived for pointer to member functions. The previous example can be transformed to the following (still using the same base class device):

```
template <typename Command, void (Command::*action)()>
struct smart_device : device
{
    smart_device(Command& command) : cmd(command) {}

    void output() override
    {
        (cmd.*action)();
    }
private:
    Command& cmd;
};


struct hello_command
{
    void say_hello_in_english()
    {
        std::cout << "Hello, world!\n";
    }

    void say_hello_in_spanish()
    {
        std::cout << "Hola mundo!\n";
    }
};
```

These classes can be used as follows:

```
hello_command cmd;

auto w1 = std::make_unique<
    smart_device<hello_command,
        &hello_command::say_hello_in_english>>(cmd);
w1->output();

auto w2 = std::make_unique<
    smart_device<hello_command,
        &hello_command::say_hello_in_spanish>>(cmd);
w2->output();
```

In C++17, a new form of specifying non-type template parameters was introduced, using the `auto` specifier (including the `auto*` and `auto&` forms) or `decltype(auto)` instead of the name of the type. This allows the compiler to deduce the type of the parameter from the expression supplied as the argument. If the deduced type is not permitted for a non-type template parameter the compiler will generate an error. Let's see an example:

```
template <auto x>
struct foo
{ /* ... */ };
```

This class template can be used as follows:

```
foo<42>   f1;   // foo<int>
foo<42.0> f2;   // foo<double> in C++20, error for older
                // versions
foo<"42"> f3;   // error
```

In the first example, for `f1`, the compiler deduces the type of the argument as `int`. In the second example, for `f2`, the compiler deduces the type as `double`. However, this is only the case for C++20. In previous versions of the standard, this line would yield an error, since floating-point types were not permitted as arguments for non-type template parameters prior to C++20. The last line, however, produces an error because `"42"` is a string literal and string literals cannot be used as arguments for non-type template parameters.

The last example can be, however, worked around in C++20 by wrapping the literal string in a structural literal class. This class would store the characters of the string literal in a fixed-length array. This is exemplified in the following snippet:

```
template<size_t N>
struct string_literal
{
    constexpr string_literal(const char(&str)[N])
    {
        std::copy_n(str, N, value);
    }

    char value[N];
};
```

However, the foo class template shown previously needs to be modified to use string_literal explicitly and not the auto specifier:

```
template <string_literal x>
struct foo
{
};
```

With this is in place, the foo<"42"> f; declaration shown earlier will compile without any errors in C++20.

The auto specifier can also be used with a non-type template parameter pack. In this case, the type is deduced independently for each template argument. The types of the template arguments do not need to be the same. This is shown in the following snippet:

```
template<auto... x>
struct foo
{ /* ... */ };

foo<42, 42.0, false, 'x'> f;
```

In this example, the compiler deduces the types of the template arguments as int, double, bool, and char, respectively.

The third and last category of template parameters are **template template parameters**. We will look at them next.

Template template parameters

Although the name may sound a bit strange, it refers to a category of template parameters that are themselves templates. These can be specified similarly to type template parameters, with or without a name, with or without a default value, and as a parameter pack with or without a name. As of C++17, both the keywords class and typename can be used to introduce a template template parameter. Prior to this version, only the class keyword could be used.

To showcase the use of template template parameters, let's consider the following two class templates first:

```cpp
template <typename T>
class simple_wrapper
{
public:
    T value;
};


template <typename T>
class fancy_wrapper
{
public:
    fancy_wrapper(T const v) :value(v)
    {
    }

    T const& get() const { return value; }

    template <typename U>
    U as() const
    {
        return static_cast<U>(value);
    }
private:
    T value;
};
```

The `simple_wrapper` class is a very simple class template that holds a value of the type template parameter T. On the other hand, `fancy_wrapper` is a more complex wrapper implementation that hides the wrapped value and exposes member functions for data access. Next, we implement a class template called `wrapping_pair` that contains two values of a wrapping type. This can be either `simpler_wrapper`, `fancy_wrapper`, or anything else that is similar:

```
template <typename T, typename U,
          template<typename> typename W = fancy_wrapper>
class wrapping_pair
{
public:
    wrapping_pair(T const a, U const b) :
        item1(a), item2(b)
    {
    }

    W<T> item1;
    W<U> item2;
};
```

The `wrapping_pair` class template has three parameters. The first two are type template parameters, named T and U. The third parameter is a template template parameter, called W, that has a default value, which is the `fancy_wrapper` type. We can use this class template as shown in the following snippet:

```
wrapping_pair<int, double> p1(42, 42.0);
std::cout << p1.item1.get() << ' '
          << p1.item2.get() << '\n';

wrapping_pair<int, double, simple_wrapper> p2(42, 42.0);
std::cout << p2.item1.value << ' '
          << p2.item2.value << '\n';
```

In this example, `p1` is a `wrapping_pair` object that contains two values, an `int` and a `double`, each wrapped in a `fancy_wrapper` object. This is not explicitly specified but is the default value of the template template parameter. On the other hand, `p2` is also a `wrapping_pair` object, also containing an `int` and a `double`, but these are wrapped by a `simple_wrapper` object, which is now specified explicitly in the template instantiation.

In this example, we have seen the use of a default template argument for a template parameter. This topic is explored in detail in the next section.

Default template arguments

Default template arguments are specified similarly to default function arguments, in the parameter list after the equal sign. The following rules apply to default template arguments:

- They can be used with any kind of template parameters with the exception of parameter packs.

- If a default value is specified for a template parameter of a class template, variable template, or type alias, then all subsequent template parameters must also have a default value. The exception is the last parameter if it is a template parameter pack.

- If a default value is specified for a template parameter in a function template, then subsequent template parameters are not restricted to also have a default value.

- In a function template, a parameter pack may be followed by more type parameters only if they have default arguments or their value can be deduced by the compiler from the function arguments.

- They are not allowed in declarations of friend class templates.

- They are allowed in the declaration of a friend function template only if the declaration is also a definition and there is no other declaration of the function in the same translation unit.

- They are not allowed in the declaration or definition of an explicit specialization of a function template or member function template.

The following snippet shows examples for using default template arguments:

```
template <typename T = int>
class foo { /*...*/ };


template <typename T = int, typename U = double>
class bar { /*...*/ };
```

As mentioned previously, a template parameter with a default argument cannot be followed by parameters without a default argument when declaring a class template but this restriction does not apply to function templates. This is shown in the next snippet:

```
template <typename T = int, typename U>
class bar { };    // error


template <typename T = int, typename U>
void func() {}    // OK
```

A template may have multiple declarations (but only one definition). The default template arguments from all the declarations and the definition are merged (the same way they are merged for default function arguments). Let's look at an example to understand how it works:

```
template <typename T, typename U = double>
struct foo;


template <typename T = int, typename U>
struct foo;


template <typename T, typename U>
struct foo
{
    T a;
    U b;
};
```

This is semantically equivalent to the following definition:

```
template <typename T = int, typename U = double>
struct foo
{
    T a;
    U b;
};
```

However, these multiple declarations with different default template arguments cannot be provided in any order. The rules mentioned earlier still apply. Therefore, a declaration of a class template where the first parameter has a default argument and the ensuing parameters do not have one is illegal:

```
template <typename T = int, typename U>
struct foo;   // error, U does not have a default argument


template <typename T, typename U = double>
struct foo;
```

Another restriction on default template arguments is that the same template parameter cannot be given multiple defaults in the same scope. Therefore, the next example will produce an error:

```
template <typename T = int>
struct foo;


template <typename T = int> // error redefinition
                            // of default parameter
struct foo {};
```

When a default template argument uses names from a class, the member access restrictions are checked at the declaration, not at the instantiation of the template:

```
template <typename T>
struct foo
{
protected:
    using value_type = T;
};


template <typename T, typename U = typename T::value_type>
struct bar
{
    using value_type = U;
};


bar<foo<int>> x;
```

When the x variable is defined, the bar class template is instantiated, but the `foo::value_type` typedef is protected and therefore cannot be used outside of `foo`. The result is a compiler error at the declaration of the `bar` class template.

With these mentions, we wrap up the topic of template parameters. The next one we will explore in the following section is template instantiation, which is the creation of a new definition of a function, class, or variable from a template definition and a set of template arguments.

Understanding template instantiation

As mentioned before, templates are only blueprints from which the compiler creates actual code when it encounters their use. The act of creating a definition for a function, a class, or a variable from the template declaration is called **template instantiation**. This can be either **explicit**, when you tell the compiler when it should generate a definition, or **implicit**, when the compiler generates a new definition as needed. We will look at these two forms in detail in the next sections.

Implicit instantiation

Implicit instantiation occurs when the compiler generates definitions based on the use of templates and when no explicit instantiation is present. Implicitly instantiated templates are defined in the same namespace as the template. However, the way compilers create definitions from templates may differ. This is something we will see in the following example. Let's consider this code:

```
template <typename T>
struct foo
{
  void f() {}
};

int main()
{
  foo<int> x;
}
```

Here, we have a class template called `foo` with a member function `f`. In `main`, we define a variable of the type `foo<int>` but do not use any of its members. Because it encounters this use of `foo`, the compiler implicitly defines a specialization of `foo` for the `int` type. If you use `cppinsights.io`, which runs in Clang, you will see the following code:

```
template<>
struct foo<int>
{
    inline void f();
};
```

Because the function `f` is not invoked in our code, it is only declared but not defined. Should we add a call `f` in `main`, the specialization would change as follows:

```
template<>
struct foo<int>
{
    inline void f() { }
};
```

However, if we add one more function, `g`, with the following implementation that contains an error, we will get different behaviors with different compilers:

```
template <typename T>
struct foo
{
    void f() {}
    void g() {int a = "42";}
};

int main()
{
    foo<int> x;
    x.f();
}
```

The body of g contains an error (you could also use a `static_assert(false)` statement as an alternative). This code compiles without any problem with VC++, but fails with Clang and GCC. This is because VC++ ignores the parts of the template that are not used, provided that the code is syntactically correct, but the others perform semantic validation before proceeding with template instantiation.

For function templates, implicit instantiation occurs when the user code refers to a function in a context that requires its definition to exist. For class templates, implicit instantiation occurs when the user code refers to a template in a context when a complete type is required or when the completeness of the type affects the code. The typical example of such a context is when an object of such a type is constructed. However, this is not the case when declaring pointers to a class template. To understand how this works, let's consider the following example:

```
template <typename T>
struct foo
{
  void f() {}
  void g() {}
};

int main()
{
   foo<int>* p;
   foo<int> x;
   foo<double>* q;
}
```

In this snippet, we use the same `foo` class template from the previous examples, and we declare several variables: p which is a pointer to `foo<int>`, x which is a `foo<int>`, and q which is a pointer to `foo<double>`. The compiler is required to instantiate only `foo<int>` at this point because of the declaration of x. Now, let's consider some invocations of the member functions f and g as follows:

```
int main()
{
   foo<int>* p;
   foo<int> x;
   foo<double>* q;
```

```
    x.f();
    q->g();
}
```

With these changes, the compiler is required to instantiate the following:

- `foo<int>` when the x variable is declared
- `foo<int>::f()` when the `x.f()` call occurs
- `foo<double>` and `foo<double>::g()` when the `q->g()` call occurs.

On the other hand, the compiler is not required to instantiate `foo<int>` when the p pointer is declared nor `foo<double>` when the q pointer is declared. However, the compiler does need to implicitly instantiate a class template specialization when it is involved in pointer conversion. This is shown in the following example:

```
template <typename T>
struct control
{};

template <typename T>
struct button : public control<T>
{};

void show(button<int>* ptr)
{
    control<int>* c = ptr;
}
```

In the function `show`, a conversion between `button<int>*` and `control<int>*` takes place. Therefore, at this point, the compiler must instantiate `button<int>`.

When a class template contains static members, those members are not implicitly instantiated when the compiler implicitly instantiates the class template but only when the compiler needs their definition. On the other hand, every specialization of a class template has its own copy of static members as exemplified in the following snippet:

```
template <typename T>
struct foo
{
    static T data;
```

```
};

template <typename T> T foo<T>::data = 0;

int main()
{
    foo<int> a;
    foo<double> b;
    foo<double> c;

    std::cout << a.data << '\n'; // 0
    std::cout << b.data << '\n'; // 0
    std::cout << c.data << '\n'; // 0

    b.data = 42;
    std::cout << a.data << '\n'; // 0
    std::cout << b.data << '\n'; // 42
    std::cout << c.data << '\n'; // 42
}
```

The class template foo has a static member variable called data that is initialized after the definition of foo. In the main function, we declare the variable a as an object of foo<int> and b and c as objects of foo<double>. Initially, all of them have the member field data initialized with 0. However, the variables b and c share the same copy of data. Therefore, after the assignment b.data = 42, a.data is still 0, but both b.data and c.data are 42.

Having learned how implicit instantiation works, it is time to move forward and understand the other form of template instantiation, which is explicit instantiation.

Explicit instantiation

As a user, you can explicitly tell the compiler to instantiate a class template or a function template. This is called explicit instantiation and it has two forms: **explicit instantiation definition** and **explicit instantiation declaration**. We will discuss them in this order.

Explicit instantiation definition

An explicit instantiation definition may appear anywhere in a program but after the definition of the template it refers to. The syntax for explicit template instantiation definitions takes the following forms:

- The syntax for class templates is as follows:

```
template class-key template-name <argument-list>
```

- The syntax for function templates is as follows:

```
template return-type name<argument-list>(parameter-list);
template return-type name(parameter-list);
```

As you can see, in all cases, the explicit instantiation definition is introduced with the `template` keyword but not followed by any parameter list. For class templates, the `class-key` can be any of the `class`, `struct`, or `union` keywords. For both class and function templates, an explicit instantiation definition with a given argument list can only appear once in the entire program.

We will look at some examples to understand how this works. Here is the first example:

```
namespace ns
{
    template <typename T>
    struct wrapper
    {
        T value;
    };

    template struct wrapper<int>;        // [1]
}

template struct ns::wrapper<double>;    // [2]

int main() {}
```

In this snippet, `wrapper<T>` is a class template defined in the `ns` namespace. The statements marked with `[1]` and `[2]` in the code are both representing an explicit instantiation definition, for `wrapper<int>` and `wrapper<double>` respectively. An explicit instantiation definition can only appear in the same namespace as the template it refers to (as in `[1]`) to or it must be fully qualified (as in `[2]`). We can write similar explicit template definitions for a function template:

```
namespace ns
{
    template <typename T>
    T add(T const a, T const b)
    {
        return a + b;
    }

    template int add(int, int);              // [1]
}

template double ns::add(double, double); // [2]

int main() { }
```

This second example has a striking resemblance to the first. Both `[1]` and `[2]` represent explicit template definitions for `add<int>()` and `add<double>()`.

If the explicit instantiation definition is not in the same namespace as the template, the name must be fully qualified. The use of a `using` statement does not make the name visible in the current namespace. This is shown in the following example:

```
namespace ns
{
    template <typename T>
    struct wrapper { T value; };
}

using namespace ns;

template struct wrapper<double>;    // error
```

The last line in this example generates a compile error because `wrapper` is an unknown name and must be qualified with the namespace name, as in `ns::wrapper`.

When class members are used for return types or parameter types, member access specification is ignored in explicit instantiation definitions. An example is shown in the following snippet:

```
template <typename T>
class foo
{
    struct bar {};

    T f(bar const arg)
    {
        return {};
    }
};


template int foo<int>::f(foo<int>::bar);
```

Both the class `X<T>::bar` and the function `foo<T>::f()` are private to the `foo<T>` class, but they can be used in the explicit instantiation definition shown on the last line.

Having seen what explicit instantiation definition is and how it works, the question that arises is when is it useful. Why would you tell the compiler to generate instantiation from a template? The answer is that it helps distribute libraries, reduce build times, and executable sizes. If you are building a library that you want to distribute as a `.lib` file and that library uses templates, the template definitions that are not instantiated are not put into the library. But that leads to increased build times of your user code every time you use the library. By forcing instantiations of templates in the library, those definitions are put into the object files and the `.lib` file you are distributing. As a result, your user code only needs to be linked to those available functions in the library file. This is what the Microsoft MSVC CRT libraries do for all the stream, locale, and string classes. The `libstdc++` library does the same for string classes and others.

A problem that can arise with template instantiations is that you can end up with multiple definitions, one per translation unit. If the same header that contains a template is included in multiple translation units (`.cpp` files) and the same template instantiation is used (let's say `wrapper<int>` from our previous examples), then identical copies of these instantiations are put in each translation unit. This leads to increased object sizes. The problem can be solved with the help of explicit instantiation declarations, which we will look at next.

Explicit instantiation declaration

An explicit instantiation declaration (available with C++11) is the way you can tell the compiler that the definition of a template instantiation is found in a different translation unit and that a new definition should not be generated. The syntax is the same as for explicit instantiation definitions except that the keyword `extern` is used in front of the declaration:

- The syntax for class templates is as follows:

```
extern template class-key template-name <argument-list>
```

- The syntax for function templates is as follows:

```
extern template return-type name<argument-
list>(parameter-list);
extern template return-type name(parameter-list);
```

If you provide an explicit instantiation declaration but no instantiation definition exists in any translation unit of the program, then the result is a compiler warning and a linker error. The technique is to declare an explicit template instantiation in one source file and explicit template declarations in the remaining ones. This will reduce both compilation times and object file sizes.

Let's look at the following example:

```
// wrapper.h
template <typename T>
struct wrapper
{
    T data;
};

extern template wrapper<int>;     // [1]

// source1.cpp
#include "wrapper.h"
#include <iostream>

template wrapper<int>;            // [2]
```

```
void f()
{
    ext::wrapper<int> a{ 42 };
    std::cout << a.data << '\n';
}


// source2.cpp
#include "wrapper.h"
#include <iostream>

void g()
{
    wrapper<int> a{ 100 };
    std::cout << a.data << '\n';
}


// main.cpp
#include "wrapper.h"

int main()
{
    wrapper<int> a{ 0 };
}
```

In this example, we can see the following:

- The wrapper.h header contains a class template called wrapper<T>. On the line marked with [1] there is an explicit instantiation declaration for wrapper<int> that tells the compiler not to generate definitions for this instantiation when a source file (translation unit) including this header is compiled.

- The source1.cpp file includes wrapper.h and on the line marked with [2] contains an explicit instantiation definition for wrapper<int>. This is the only definition for this instantiation within the entire program.

- The source files source2.cpp and main.cpp are both using wrapper<int> but without any explicit instantiation definition or declaration. That is because the explicit declaration from wrapper.h is visible when the header is included in each of these files.

Alternatively, the explicit instantiation declaration could be taken away from the header file but then it must be added to each source file that includes the header and that is likely to be forgotten.

When you do explicit template declarations, keep in mind that a class member function that is defined within the body of the class is always considered inline and therefore it will always be instantiated. Therefore, you can only use the `extern` keyword for member functions that are defined outside of the class body.

Now that we have looked at what template instantiation is, we will continue with another important topic, **template specialization**, which is the term used for the definition created from a template instantiation to handle a specific set of template arguments.

Understanding template specialization

A **template specialization** is the definition created from a template instantiation. The template that is being specialized is called the **primary template**. You can provide an explicit specialized definition for a given set of template arguments, therefore overwriting the implicit code the compiler would generate instead. This is the technique that powers features such as type traits and conditional compilation, which are metaprogramming concepts we will explore in *Chapter 5, Type Traits and Conditional Compilation*.

There are two forms of template specialization: **explicit (full) specialization** and **partial specialization**. We will look in detail at these two in the following sections.

Explicit specialization

Explicit specialization (also called full specialization) occurs when you provide a definition for a template instantiation with the full set of template arguments. The following can be fully specialized:

- Function templates
- Class templates
- Variable templates (as of C++14)
- Member functions, classes, and enumerations of a class template
- Member function templates and class templates of a class or class template
- Static data members of a class template

Let's start by looking at the following example:

```cpp
template <typename T>
struct is_floating_point
{
    constexpr static bool value = false;
};

template <>
struct is_floating_point<float>
{
    constexpr static bool value = true;
};

template <>
struct is_floating_point<double>
{
    constexpr static bool value = true;
};

template <>
struct is_floating_point<long double>
{
    constexpr static bool value = true;
};
```

In this code snippet, `is_floating_point` is the primary template. It contains a `constexpr` static Boolean data member called `value` that is initialized with the `false` value. Then, we have three full specializations of this primary template, for the `float`, `double`, and `long double` types. These new definitions change the way `value` is being initialized using `true` instead of `false`. As a result, we can use this template to write code as follows:

```cpp
std::cout << is_floating_point<int>::value          << '\n';
std::cout << is_floating_point<float>::value        << '\n';
std::cout << is_floating_point<double>::value        << '\n';
std::cout << is_floating_point<long double>::value << '\n';
std::cout << is_floating_point<std::string>::value << '\n';
```

The first and last lines print 0 (for false); the other lines print 1 (for true). This example is a demonstration of how type traits work. In fact, the standard library contains a class template called is_floating_point in the std namespace, defined in the <type_traits> header. We will learn more about this topic in *Chapter 5, Type Traits and Conditional Compilation*.

As you can see in this example, static class members can be fully specialized. However, each specialization has its own copy of any static members, which is demonstrated with the following example:

```
template <typename T>
struct foo
{
    static T value;
};


template <typename T> T foo<T>::value = 0;
template <> int foo<int>::value = 42;


foo<double> a, b;    // a.value=0, b.value=0
foo<int> c;          // c.value=42


a.value = 100;       // a.value=100, b.value=100, c.value=42
```

Here, foo<T> is a class template with a single static member, called value. This is initialized with 0 for the primary template and with 42 for the int specialization. After declaring the variables a, b, and c, a.value is 0 and b.value is 0 while c.value is 42. However, after assigning the value 100 to a.value, b.value is also 100, while c.value remains 42.

Explicit specialization must appear after the primary template declaration. It does not require a definition of the primary template to be available prior to the explicit specialization. The following code is therefore valid:

```
template <typename T>
struct is_floating_point;


template <>
struct is_floating_point<float>
{
```

```
    constexpr static bool value = true;
};

template <typename T>
struct is_floating_point
{
    constexpr static bool value = false;
};
```

Template specializations can also be only declared without being defined. Such a template specialization can be used like any other incomplete type. You can see an example of this here:

```
template <typename>.
struct foo {};      // primary template

template <>
struct foo<int>;   // explicit specialization declaration

foo<double> a; // OK
foo<int>* b;      // OK
foo<int> c;       // error, foo<int> incomplete type
```

In this example, foo<T> is the primary template for which a declaration of an explicit specialization for the int type exists. This makes it possible to use foo<double> and foo<int>* (declaring pointers to partial types is supported). However, at the point of declaring the c variable, the complete type foo<int> is not available, since a definition of the full specialization for int is missing. This generates a compiler error.

When specializing a function template, if the compiler can deduce a template argument from the type of the function arguments, then that template argument is optional. This is demonstrated by the following example:

```
template <typename T>
struct foo {};

template <typename T>
void func(foo<T>)
{
```

```
        std::cout << "primary template\n";
}


template<>
void func(foo<int>)
{
    std::cout << "int specialization\n";
}
```

The syntax for the full specialization for `int` of the `func` function template should be `template<> func<int>(foo<int>)`. However, the compiler is able to deduce the actual type that `T` represents from the function argument. Therefore, we don't have to specify it when defining the specialization.

On the other hand, declarations or definitions of function templates and member function templates are not allowed to contain default function arguments. Therefore, in the following example, the compiler will issue an error:

```
template <typename T>
void func(T a)
{
    std::cout << "primary template\n";
}


template <>
void func(int a = 0)  // error: default argument not allowed
{
    std::cout << "int specialization\n";
}
```

In all these examples, the templates had a single template argument. However, in practice, many templates have multiple arguments. Explicit specialization requires a definition with the full set of arguments being specified. This is demonstrated with the following snippet:

```
template <typename T, typename U>
void func(T a, U b)
{
    std::cout << "primary template\n";
```

```
}

template <>
void func(int a, int b)
{
    std::cout << "int-int specialization\n";
}


template <>
void func(int a, double b)
{
    std::cout << "int-double specialization\n";
}


func(1, 2);        // int-int specialization
func(1, 2.0);      // int-double specialization
func(1.0, 2.0);    // primary template
```

With these covered, we can move forward and look into partial specialization, which is basically a generalization of explicit (full) specialization.

Partial specialization

Partial specialization occurs when you specialize a primary template but only specify some of the template arguments. This means a partial specialization has both a template parameter list (which follows the template keyword) and a template argument list (which follows the template name). However, only classes can be partially specialized.

Let's explore the following example to understand how this works:

```
template <typename T, int S>
struct collection
{
    void operator()()
    { std::cout << "primary template\n"; }
};
```

```cpp
template <typename T>
struct collection<T, 10>
{
    void operator()()
    { std::cout << "partial specialization <T, 10>\n"; }
};

template <int S>
struct collection<int, S>
{
    void operator()()
    { std::cout << "partial specialization <int, S>\n"; }
};

template <typename T, int S>
struct collection<T*, S>
{
    void operator()()
    { std::cout << "partial specialization <T*, S>\n"; }
};
```

We have a primary template called `collection` that has two template arguments (a type template argument and a non-type template argument) and we have three partial specializations, as follows:

- A specialization for the non-type template argument `S` with the value `10`
- A specialization for the `int` type
- A specialization for the pointer type `T*`

These templates can be used as shown in the following snippet:

```cpp
collection<char, 42> a;   // primary template
collection<int,  42> b;   // partial specialization <int, S>
collection<char, 10> c;   // partial specialization <T, 10>
collection<int*, 20> d;   // partial specialization <T*, S>
```

As specified in the comments, a is instantiated from the primary template, b from the partial specialization for int (collection<int, S>), c from the partial specialization for 10 (collection<T, 10>), and d from the partial specialization for pointers (collection<T*, S>). However, some combinations are not possible because they are ambiguous and the compiler cannot select which template instantiation to use. Here are a couple of examples:

```
collection<int,    10> e; // error: collection<T,10> or
                      //           collection<int,S>
collection<char*, 10> f; // error: collection<T,10> or
                      //           collection<T*,S>
```

In the first case, both collection<T, 10> and collection<int, S> partial specializations match the type collection<int, 10>, while in the second case it can be either collection<T, 10> or collection<T*, S>.

When defining specializations of a primary template, you need to keep in mind the following:

- Parameters in the template parameters list of the partial specialization cannot have default values.
- The template parameters list implies an order of the arguments in the template arguments list, which is featured only in a partial specialization. This template arguments list of a partial specialization cannot be the same as the one implied by the template parameters list.
- In the template argument list, you can only use identifiers for non-type template parameters. Expressions are not allowed in this context. This is demonstrated with the following example:

```
template <int A, int B> struct foo {};
template <int A> struct foo<A, A> {};        // OK
template <int A> struct foo<A, A + 1> {}; // error
```

When a class template has partial specializations, the compiler must decide what is the best match to generate a definition from. For this purpose, it matches the template arguments of the template specialization with the template argument list of the primary template and partial specializations. Depending on the result of this matching process, the compiler does the following:

- If no match is found, a definition is generated from the primary template.
- If a single partial specialization is found, a definition is generated from that specialization.

- If more than a single partial specialization is found, then a definition is generated from the most specialized partial specialization but only if it is unique. Otherwise, the compiler generates an error (as we have seen previously). A template A is considered more specialized than a template B if it accepts a subset of the types that B accepts, but not the other way around.

However, partial specializations are not found by name lookup and are considered only if the primary template is found by name lookup.

To understand how partial specialization is useful, let's take a look at a real-world example.

In this example, we want to create a function that formats the content of an array in a nice way and outputs it to a stream. The content of a formatted array should look like *[1,2,3,4,5]*. However, for arrays of char elements, the elements should not be separated by a comma but instead displayed as a string within square brackets, such as *[demo]*. For this purpose, we will consider the use of the std::array class. The following implementation formats the content of the array with delimiters between the elements:

```
template <typename T, size_t S>
std::ostream& pretty_print(std::ostream& os,
                           std::array<T, S> const& arr)
{
    os << '[';
    if (S > 0)
    {
        size_t i = 0;
        for (; i < S - 1; ++i)
            os << arr[i] << ',';
        os << arr[S-1];
    }
    os << ']';

    return os;
}

std::array<int, 9> arr {1, 1, 2, 3, 5, 8, 13, 21};
pretty_print(std::cout, arr);   // [1,1,2,3,5,8,13,21]
```

```
std::array<char, 9> str;
std::strcpy(str.data(), "template");
pretty_print(std::cout, str);   // [t,e,m,p,l,a,t,e]
```

In this snippet, pretty_print is a function template with two template parameters, matching the template parameters of the std::array class. When called with the arr array as an argument, it prints *[1,1,2,3,5,8,13,21]*. When called with the str array as an argument, it prints *[t,e,m,p,l,a,t,e]*. However, our intention is to print *[template]* in this latter case. For this, we need another implementation, which is specialized for the char type:

```
template <size_t S>
std::ostream& pretty_print(std::ostream& os,
                           std::array<char, S> const& arr)
{
    os << '[';
    for (auto const& e : arr)
        os << e;
    os << ']';

    return os;
}

std::array<char, 9> str;
std::strcpy(str.data(), "template");
pretty_print(std::cout, str);   // [template]
```

In this second implementation, pretty_print is a function template with a single template parameter, which is a non-type template parameter indicating the size of the array. The type template parameter is explicitly specified as char, in std::array<char, S>. This time, the call to pretty_print with the str array prints [template] to the console.

What is key to understand here is that it's not the pretty_print function template that is partially specialized but the std::array class template. Function templates cannot be specialized and what we have here are overloaded functions. However, std::array<char,S> is a specialization of the primary class template std::array<T, S>.

All the examples we have seen in this chapter were either function templates or class templates. However, variables can also be templates and this will be the topic of the next section.

Defining variable templates

Variable templates were introduced in C++14 and allow us to define variables that are templates either at namespace scope, in which case they represent a family of global variables, or at class scope, in which case they represent static data members.

A variable template is declared at a namespace scope as shown in the following code snippet. This is a typical example that you can find in the literature, but we can use it to elaborate on the benefits of variable templates:

```
template<class T>
constexpr T PI = T(3.1415926535897932385L);
```

The syntax is similar to declaring a variable (or data member) but combined with the syntax for declaring templates.

The question that arises is how variable templates are actually helpful. To answer this, let's build up an example to demonstrate the point. Let's consider we want to write a function template that, given the radius of a sphere, returns its volume. The volume of a sphere is $4\pi r^3 / 3$. Therefore, a possible implementation is as follows:

```
constexpr double PI = 3.1415926535897932385L;

template <typename T>
T sphere_volume(T const r)
{
    return 4 * PI * r * r * r / 3;
}
```

In this example, PI is defined as a compile-time constant of the double type. This will generate a compiler warning if we use float, for instance, for the type template parameter T:

```
float v1 = sphere_volume(42.0f); // warning
double v2 = sphere_volume(42.0); // OK
```

A potential solution to this problem is to make `PI` a static data member of a template class with its type determined by the type template parameter. This implementation can look as follows:

```
template <typename T>
struct PI
{
    static const T value;
};

template <typename T>
const T PI<T>::value = T(3.1415926535897932385L);

template <typename T>
T sphere_volume(T const r)
{
    return 4 * PI<T>::value * r * r * r / 3;
}
```

This works, although the use of `PI<T>::value` is not ideal. It would be nicer if we could simply write `PI<T>`. This is exactly what the variable template `PI` shown at the beginning of the section allows us to do. Here it is again, with the complete solution:

```
template<class T>
constexpr T PI = T(3.1415926535897932385L);

template <typename T>
T sphere_volume(T const r)
{
    return 4 * PI<T> * r * r * r / 3;
}
```

The next example shows yet another possible use case and also demonstrates the explicit specialization of variable templates:

```
template<typename T>
constexpr T SEPARATOR = '\n';

template<>
```

```
constexpr wchar_t SEPARATOR<wchar_t> = L'\n';

template <typename T>
std::basic_ostream<T>& show_parts(
    std::basic_ostream<T>& s,
    std::basic_string_view<T> const& str)
{
    using size_type =
        typename std::basic_string_view<T>::size_type;
    size_type start = 0;
    size_type end;
    do
    {
        end = str.find(SEPARATOR<T>, start);
        s << '[' << str.substr(start, end - start) << ']'
          << SEPARATOR<T>;
        start = end+1;
    } while (end != std::string::npos);

    return s;
}

show_parts<char>(std::cout, "one\ntwo\nthree");
show_parts<wchar_t>(std::wcout, L"one line");
```

In this example, we have a function template called `show_parts` that processes an input string after splitting it into parts delimited by a separator. The separator is a variable template defined at (global) namespace scope and is explicitly specialized for the `wchar_t` type.

As previously mentioned, variable templates can be members of classes. In this case, they represent static data members and need to be declared using the `static` keyword. The following example demonstrates this:

```
struct math_constants
{
    template<class T>
    static constexpr T PI = T(3.1415926535897932385L);
```

```
};

template <typename T>
T sphere_volume(T const r)
{
    return 4 * math_constants::PI<T> *r * r * r / 3;
}
```

You can declare a variable template in a class and then provide its definition outside the class. Notice that in this case, the variable template must be declared with `static const` and not `static constexpr`, since the latter one requires in-class initialization:

```
struct math_constants
{
    template<class T>
    static const T PI;
};

template<class T>
const T math_constants::PI = T(3.1415926535897932385L);
```

Variable templates are used to simplify the use of type traits. The *Explicit specialization* section contained an example for a type trait called `is_floating_point`. Here is, again, the primary template:

```
template <typename T>
struct is_floating_point
{
    constexpr static bool value = false;
};
```

There were several explicit specializations that I will not list here again. However, this type trait can be used as follows:

```
std::cout << is_floating_point<float>::value << '\n';
```

The use of `is_floating_point<float>::value` is rather cumbersome, but can be avoided with the help of a variable template that can be defined as follows:

```
template <typename T>
inline constexpr bool is_floating_point_v =
    is_floating_point<T>::value;
```

This `is_floating_point_v` variable template helps write code that is arguably simpler and easier to read. The following snippet is the form I prefer over the verbose variant with `::value`:

```
std::cout << is_floating_point_v<float> << '\n';
```

The standard library defines a series of variable templates suffixed with `_v` for `::value`, just as in our example (such as `std::is_floating_point_v` or `std::is_same_v`). We will discuss this topic in more detail in *Chapter 5, Type Traits and Conditional Compilation.*

Variable templates are instantiated similarly to function templates and class templates. This happens either with an explicit instantiation or explicit specialization, or implicitly by the compiler. The compiler generates a definition when the variable template is used in a context where a variable definition must exist, or the variable is needed for constant evaluation of an expression.

After this, we move to the topic of alias templates, which allow us to define aliases for class templates.

Defining alias templates

In C++, an **alias** is a name used to refer to a type that has been previously defined, whether a built-in type or a user-defined type. The primary purpose of aliases is to give shorter names to types that have a long name or provide semantically meaningful names for some types. This can be done either with a `typedef` declaration or with a `using` declaration (the latter was introduced in C++11). Here are several examples using `typedef`:

```
typedef int index_t;
typedef std::vector<
            std::pair<int, std::string>> NameValueList;
typedef int (*fn_ptr)(int, char);

template <typename T>
struct foo
```

```
{
    typedef T value_type;
};
```

In this example, `index_t` is an alias for `int`, `NameValueList` is an alias for
`std::vector<std::pair<int, std::string>>`, while `fn_ptr` is an alias for
the type of a pointer to a function that returns an `int` and has two parameters of type
`int` and `char`. Lastly, `foo::value_type` is an alias for the type template `T`.

Since C++11, these type aliases can be created with the help of **using declarations**,
which look as follows:

```
using index_t = int;
using NameValueList =
    std::vector<std::pair<int, std::string>>;
using fn_ptr = int(*)(int, char);

template <typename T>
struct foo
{
    using value_type = T;
};
```

Using declarations are now preferred over `typedef` declarations because they are
simpler to use and are also more natural to read (from left to right). However, they have
an important advantage over `typedef`s as they allow us to create aliases for templates.
An **alias template** is a name that refers not to a type but a family of types. Remember,
a template is not a class, function, or variable but a blueprint that allows the creation of
a family of types, functions, or variables.

To understand how alias templates work, let's consider the following example:

```
template <typename T>
using customer_addresses_t =
    std::map<int, std::vector<T>>;                  // [1]

struct delivery_address_t {};
struct invoice_address_t {};
```

```
using customer_delivery_addresses_t =
    customer_addresses_t<delivery_address_t>; // [2]
using customer_invoice_addresses_t =
    customer_addresses_t<invoice_address_t>;  // [3]
```

The declaration on line [1] introduces the alias template `customer_addresses_t`. It's an alias for a map type where the key type is `int` and the value type is `std::vector<T>`. Since `std::vector<T>` is not a type, but a family of types, `customer_addresses_t<T>` defines a family of types. The `using` declarations at [2] and [3] introduce two type aliases, `customer_delivery_addresses_t` and `customer_invoice_addresses_t`, from the aforementioned family of types.

Alias templates can appear at namespace or class scope just like any template declaration. On the other hand, they can neither be fully nor partially specialized. However, there are ways to overcome this limitation. A solution is to create a class template with a type alias member and specialize the class. Then you can create an alias template that refers to the type alias member. Let's demonstrate this with the help of an example.

Although the following is not valid C++ code, it represents the end goal I want to achieve, had the specialization of alias templates been possible:

```
template <typename T, size_t S>
using list_t = std::vector<T>;


template <typename T>
using list_t<T, 1> = T;
```

In this example, `list_t` is an alias template for `std::vector<T>` provided the size of the collection is greater than 1. However, if there is a single element, then `list_t` should be an alias for the type template parameter `T`. The way this can be actually achieved is shown in the following snippet:

```
template <typename T, size_t S>
struct list
{
    using type = std::vector<T>;
};


template <typename T>
struct list<T, 1>
{
```

```
    using type = T;
};


template <typename T, size_t S>
using list_t = typename list<T, S>::type;
```

In this example, `list<T,S>` is a class template that has a member type alias called T. In the primary template, this is an alias for `std::vector<T>`. In the partial specialization `list<T,1>` it's an alias for T. Then, `list_t` is defined as an alias template for `list<T, S>::type`. The following asserts prove this mechanism works:

```
static_assert(std::is_same_v<list_t<int, 1>, int>);
static_assert(std::is_same_v<list_t<int, 2>,
std::vector<int>>);
```

Before we end this chapter, there is one more topic that needs to be addressed: generic lambdas and their C++20 improvement, lambda templates.

Exploring generic lambdas and lambda templates

Lambdas, which are formally called **lambda expressions**, are a simplified way to define function objects in the place where they are needed. This typically includes predicates or comparison functions passed to algorithms. Although we will not discuss lambda expressions in general, let's take a look at the following examples:

```
int arr[] = { 1,6,3,8,4,2,9 };
std::sort(
    std::begin(arr), std::end(arr),
    [](int const a, int const b) {return a > b; });

int pivot = 5;
auto count = std::count_if(
    std::begin(arr), std::end(arr),
    [pivot](int const a) {return a > pivot; });
```

Lambda expressions are syntactic sugar, a simplified way of defining anonymous function objects. When encountering a lambda expression, the compiler generates a class with a function-call operator. For the previous example, these could look as follows:

```
struct __lambda_1
{
    inline bool operator()(const int a, const int b) const
    {
        return a > b;
    }
};

struct __lambda_2
{
    __lambda_2(int & _pivot) : pivot{_pivot}
    {}

    inline bool operator()(const int a) const
    {
        return a > pivot;
    }
private:
    int pivot;
};
```

The names chosen here are arbitrary and each compiler will generate different names. Also, the implementation details may differ and the ones seen here are the bare minimum a compiler is supposed to generate. Notice that the difference between the first lambda and the second is that the latter contains state that it captures by value.

Lambda expressions, which were introduced in C++11, have received several updates in later versions of the standard. There are notably two, which will be discussed in this chapter:

- **Generic lambdas**, introduced in C++14, allow us to use the `auto` specifier instead of explicitly specifying types. This transforms the generated function object into one with a template function-call operator.

- **Template lambdas**, introduced in C++20, allow us to use the template syntax to explicitly specify the shape of the templatized function-call operator.

To understand the difference between these and how generic and template lambdas are helpful, let's explore the following examples:

```
auto l1 = [] (int a) {return a + a; };   // C++11, regular
                                         // lambda
auto l2 = [] (auto a) {return a + a; }; // C++14, generic
                                         // lambda
auto l3 = []<typename T>(T a)
          { return a + a; };     // C++20, template lambda

auto v1 = l1(42);                        // OK
auto v2 = l1(42.0);                      // warning
auto v3 = l1(std::string{ "42" });       // error

auto v5 = l2(42);                        // OK
auto v6 = l2(42.0);                      // OK
auto v7 = l2(std::string{"42"});         // OK

auto v8 = l3(42);                        // OK
auto v9 = l3(42.0);                      // OK
auto v10 = l3(std::string{ "42" });      // OK
```

Here, we have three different lambdas: l1 is a regular lambda, l2 is a generic lambda, as at least one of the parameters is defined with the auto specifier, and l3 is a template lambda, defined with the template syntax but without the use of the template keyword.

We can invoke l1 with an integer; we can also invoke it with a double, but this time the compiler will produce a warning about the possible loss of data. However, trying to invoke it with a string argument will produce a compile error, because std::string cannot be converted to int. On the other hand, l2 is a generic lambda. The compiler proceeds to instantiate specializations of it for all the types of the arguments it's invoked with, in this example int, double, and std::string. The following snippet shows how the generated function object may look, at least conceptually:

```
struct __lambda_3
{
    template<typename T1>
    inline auto operator()(T1 a) const
    {
```

```
      return a + a;
   }


   template<>
   inline int operator()(int a) const
   {
      return a + a;
   }


   template<>
   inline double operator()(double a) const
   {
      return a + a;
   }


   template<>
   inline std::string operator()(std::string a) const
   {
      return std::operator+(a, a);
   }
};
```

You can see here the primary template for the function-call operator, as well as the three specializations that we mentioned. Not surprisingly, the compiler will generate the same code for the third lambda expression, 13, which is a template lambda, only available in C++20. The question that arises from this is how are generic lambdas and lambda templates different? To answer this question, let's modify the previous example a bit:

```
auto l1 = [](int a, int b) {return a + b; };
auto l2 = [](auto a, auto b) {return a + b; };
auto l3 = []<typename T, typename U>(T a, U b)
          { return a + b; };

auto v1 = l1(42, 1);                      // OK
auto v2 = l1(42.0, 1.0);                  // warning
auto v3 = l1(std::string{ "42" }, '1');   // error
```

```
auto v4 = 12(42, 1);                          // OK
auto v5 = 12(42.0, 1);                        // OK
auto v6 = 12(std::string{ "42" }, '1'); // OK
auto v7 = 12(std::string{ "42" }, std::string{ "1" }); // OK


auto v8 = 13(42, 1);                          // OK
auto v9 = 13(42.0, 1);                        // OK
auto v10 = 13(std::string{ "42" }, '1'); // OK
auto v11 = 13(std::string{ "42" }, std::string{ "42" }); // OK
```

The new lambda expressions take two parameters. Again, we can call 11 with two integers or an int and a double (although this again generates a warning) but we can't call it with a string and char. However, we can do all these with the generic lambda 12 and the lambda template 13. The code the compiler generates is identical for 12 and 13 and looks, semantically, as follows:

```
struct __lambda_4
{
    template<typename T1, typename T2>
    inline auto operator()(T1 a, T2 b) const
    {
      return a + b;
    }


    template<>
    inline int operator()(int a, int b) const
    {
      return a + b;
    }


    template<>
    inline double operator()(double a, int b) const
    {
      return a + static_cast<double>(b);
    }
```

```
template<>
inline std::string operator()(std::string a,
                              char b) const
{
  return std::operator+(a, b);
}


template<>
inline std::string operator()(std::string a,
                              std::string b) const
{
  return std::operator+(a, b);
}
};
```

We see, in this snippet, the primary template for the function-call operator, and several full explicit specializations: for two `int` values, for a `double` and an `int`, for a string and a `char`, and for two string objects. But what if we want to restrict the use of the generic lambda 12 to arguments of the same type? This is not possible. The compiler cannot deduce our intention and, therefore, it would generate a different type template parameter for each occurrence of the `auto` specifier in the parameter list. However, the lambda templates from C++20 do allow us to specify the form of the function-call operator. Take a look at the following example:

```
auto 15 = []<typename T>(T a, T b) { return a + b; };

auto v1 = 15(42, 1);          // OK
auto v2 = 15(42, 1.0);        // error

auto v4 = 15(42.0, 1.0);      // OK
auto v5 = 15(42, false);      // error

auto v6 = 15(std::string{ "42" }, std::string{ "1" }); // OK

auto v6 = 15(std::string{ "42" }, '1'); // error
```

Invoking the lambda template with any two arguments of different types, even if they are implicitly convertible such as from int to double, is not possible. The compiler will generate an error. It's not possible to explicitly provide the template arguments when invoking the template lambda, such as in l5<double>(42, 1.0). This also generates a compiler error.

The decltype type specifier allows us to tell the compiler to deduce the type from an expression. This topic is covered in detail in *Chapter 4, Advanced Template Concepts*. However, in C++14, we can use this in a generic lambda to declare the second parameter in the previous generic lambda expression to have the same type as the first parameter. More precisely, this would look as follows:

```
auto l4 = [](auto a, decltype(a) b) {return a + b; };
```

However, this implies that the type of the second parameter, b, must be convertible to the type of the first parameter, a. This allows us to write the following calls:

```
auto v1 = l4(42.0, 1);                  // OK
auto v2 = l4(42, 1.0);                  // warning
auto v3 = l4(std::string{ "42" }, '1'); // error
```

The first call is compiled without any problems because int is implicitly convertible to double. The second call compiles with a warning, because converting from double to int may incur a loss of data. The third call, however, generates an error, because char cannot be implicitly convertible to std::string. Although the l4 lambda is an improvement over the generic lambda l2 seen previously, it still does not help restrict calls completely if the arguments are of different types. This is only possible with lambda templates as shown earlier.

Another example of a lambda template is shown in the next snippet. This lambda has a single argument, a std::array. However, the type of the elements of the array and the size of the array are specified as template parameters of the lambda template:

```
auto l = []<typename T, size_t N>(
            std::array<T, N> const& arr)
{
    return std::accumulate(arr.begin(), arr.end(),
                           static_cast<T>(0));
};

auto v1 = l(1);                        // error
auto v2 = l(std::array<int, 3>{1, 2, 3}); // OK
```

Attempting to call this lambda with anything other than an `std::array` object produces a compiler error. The compiler-generated function object may look as follows:

```
struct __lambda_5
{
    template<typename T, size_t N>
    inline auto operator()(
       const std::array<T, N> & arr) const
    {
      return std::accumulate(arr.begin(), arr.end(),
                            static_cast<T>(0));
    }

    template<>
    inline int operator()(
       const std::array<int, 3> & arr) const
    {
      return std::accumulate(arr.begin(), arr.end(),
                            static_cast<int>(0));
    }
};
```

An interesting benefit of generic lambdas over regular lambdas concerns recursive lambdas. Lambdas do not have names; they are anonymous, therefore, you cannot recursively call them directly. Instead, you have to define a `std::function` object, assign the lambda expression to it, and also capture it by reference in the capture list. The following is an example of a recursive lambda that computes the factorial of a number:

```
std::function<int(int)> factorial;
factorial = [&factorial](int const n) {
   if (n < 2) return 1;
      else return n * factorial(n - 1);
};

factorial(5);
```

This can be simplified with the use of generic lambdas. They don't require a `std::function` and its capture. A recursive generic lambda can be implemented as follows:

```
auto factorial = [](auto f, int const n) {
    if (n < 2) return 1;
    else return n * f(f, n - 1);
};


factorial(factorial, 5);
```

If understanding how this works is hard, the compiler-generated code should help you figure it out:

```
struct __lambda_6
{
    template<class T1>
    inline auto operator()(T1 f, const int n) const
    {
      if(n < 2) return 1;
      else return n * f(f, n - 1);
    }


    template<>
    inline int operator()(__lambda_6 f, const int n) const
    {
      if(n < 2) return 1;
      else return n * f.operator()(__lambda_6(f), n - 1);
    }
};


__lambda_6 factorial = __lambda_6{};
factorial(factorial, 5);
```

A generic lambda is a function object with a template function-call operator. The first argument, specified with `auto`, can be anything, including the lambda itself. Therefore, the compiler will provide a full explicit specialization of the call operator for the type of the generated class.

Lambda expressions help us avoid writing explicit code when we need to pass function objects as arguments to other functions. The compiler, instead, generates that code for us. Generic lambdas, introduced in C++14, help us avoid writing the same lambdas for different types. The lambda templates for C++20 allow us to specify the form of the generated call operator with the help of template syntax and semantics.

Summary

This chapter was a walk through of the core features of C++ templates. We have learned how to define class templates, function templates, variable templates, and alias templates. Along the way, we looked in detail at template instantiation and template specialization after learning about template parameters. We also learned about generic lambdas and lambda templates and what benefits they have compared to regular lambdas. By completing this chapter, you are now familiar with the template fundamentals, which should allow you to understand large parts of template code as well as write templates yourself.

In the next chapter, we will look at another important topic, which is templates with a variable number of arguments called variadic templates.

Questions

1. What category of types can be used for non-type template parameters?
2. Where are default template arguments not allowed?
3. What is explicit instantiation declaration and how does it differ syntactically from explicit instantiation definition?
4. What is an alias template?
5. What are template lambdas?

Further reading

- C++ Template: A Quick UpToDate Look(C++11/14/17/20), http://www.vishalchovatiya.com/c-template-a-quick-uptodate-look/
- Templates aliases for C++, https://www.stroustrup.com/template-aliases.pdf
- Lambdas: From C++11 to C++20, Part 2, https://www.cppstories.com/2019/03/lambdas-story-part2/

3
Variadic Templates

A variadic template is a template with a variable number of arguments. This is a feature that was introduced in C++11. It combines generic code with functions with variable numbers of arguments, a feature that was inherited from the C language. Although the syntax and some details could be seen as cumbersome, variadic templates help us write function templates with a variable number of arguments or class templates with a variable number of data members in a way that was not possible before with compile time evaluation and type safety.

In this chapter, we will learn about the following topics:

- Understanding the need for variadic templates
- Variadic function templates
- Parameter packs
- Variadic class templates
- Fold expressions
- Variadic alias templates
- Variadic variable templates

By the end of the chapter, you will have a good understanding of how to write variadic templates and how they work.

We will start, however, by trying to understand why templates with variable numbers of arguments are helpful.

Understanding the need for variadic templates

One of the most famous C and C++ functions is `printf`, which writes formatted output to the `stdout` standard output stream. There is actually a family of functions in the I/O library for writing formatted output, which also includes `fprintf` (which writes to a file stream), `sprint`, and `snprintf` (which write to a character buffer). These functions are similar because they take a string defining the output format and a variable number of arguments. The language, however, provides us with the means to write our own functions with variable numbers of arguments. Here is an example of a function that takes one or more arguments and returns the minimum value:

```cpp
#include<stdarg.h>

int min(int count, ...)
{
    va_list args;
    va_start(args, count);

    int val = va_arg(args, int);
    for (int i = 1; i < count; i++)
    {
        int n = va_arg(args, int);
        if (n < val)
            val = n;
    }

    va_end(args);

    return val;
}

int main()
{
    std::cout << "min(42, 7)=" << min(2, 42, 7) << '\n';
    std::cout << "min(1,5,3,-4,9)=" <<
                    min(5, 1, 5, 3, -4,
                9) << '\n';
}
```

This implementation is specific for values of the int type. However, it is possible to write a similar function that is a function template. The transformation requires minimal changes and the result is as follows:

```
template <typename T>
T min(int count, ...)
{
    va_list args;
    va_start(args, count);

    T val = va_arg(args, T);
    for (int i = 1; i < count; i++)
    {
        T n = va_arg(args, T);
        if (n < val)
            val = n;
    }

    va_end(args);

    return val;
}

int main()
{
    std::cout << "min(42.0, 7.5)="
              << min<double>(2, 42.0, 7.5) << '\n';
    std::cout << "min(1,5,3,-4,9)="
              << min<int>(5, 1, 5, 3, -4, 9) << '\n';
}
```

Writing code like this, whether generic or not, has several important drawbacks:

- It requires the use of several macros: va_list (which provides access to the information needed by the others), va_start (starts the iterating of the arguments), va_arg (provides access to the next argument), and va_end (stops the iterating of the arguments).

- Evaluation happens at runtime, even though the number and the type of the arguments passed to the function are known at compile-time.

- Variadic functions implemented in this manner are not type-safe. The `va_` macros perform low-memory manipulation and type-casts are done in `va_arg` at runtime. These could lead to runtime exceptions.

- These variadic functions require specifying in some way the number of variable arguments. In the implementation of the earlier `min` function, there is a first parameter that indicates the number of arguments. The `printf`-like functions take a formatting string from which the number of expected arguments is determined. The `printf` function, for example, evaluates and then ignores additional arguments (if more are supplied than the number specified in the formatting string) but has undefined behavior if fewer arguments are supplied.

In addition to all these things, only functions could be variadic, prior to C++11. However, there are classes that could also benefit from being able to have a variable number of data members. Typical examples are the `tuple` class, which represents a fixed-size collection of heterogeneous values, and `variant`, which is a type-safe union.

Variadic templates help address all these issues. They are evaluated at compile-time, are type-safe, do not require macros, do not require explicitly specifying the number of arguments, and we can write both variadic function templates and variadic class templates. Moreover, we also have variadic variable templates and variadic alias templates.

In the next section, we will start looking into variadic function templates.

Variadic function templates

Variadic function templates are template functions with a variable number of arguments. They borrow the use of the ellipsis (. . .) for specifying a pack of arguments, which can have different syntax depending on its nature.

To understand the fundamentals for variadic function templates, let's start with an example that rewrites the previous `min` function:

```
template <typename T>
T min(T a, T b)
{
    return a < b ? a : b;
}
```

```
template <typename T, typename... Args>
T min(T a, Args... args)
{
    return min(a, min(args...));
}

int main()
{
    std::cout << "min(42.0, 7.5)=" << min(42.0, 7.5)
              << '\n';
    std::cout << "min(1,5,3,-4,9)=" << min(1, 5, 3, -4, 9)
              << '\n';
}
```

What we have here are two overloads for the min function. The first is a function template with two parameters that returns the smallest of the two arguments. The second is a function template with a variable number of arguments that recursively calls itself with an expansion of the parameters pack. Although variadic function template implementations look like using some sort of compile-time recursion mechanism (in this case the overload with two parameters acting as the end case), in fact, they're only relying on overloaded functions, instantiated from the template and the set of provided arguments.

The ellipsis (. . .) is used in three different places, with different meanings, in the implementation of a variadic function template, as can be seen in our example:

- To specify a pack of parameters in the template parameters list, as in typename . . . Args. This is called a **template parameter pack**. Template parameter packs can be defined for type templates, non-type templates, and template template parameters.

- To specify a pack of parameters in the function parameters list, as in Args . . . args. This is called a **function parameter pack**.

- To expand a pack in the body of a function, as in args..., seen in the call min(args...). This is called a **parameter pack expansion**. The result of such an expansion is a comma-separated list of zero or more values (or expressions). This topic will be covered in more detail in the next section.

From the call min(1, 5, 3, -4, 9), the compiler is instantiating a set of overloaded functions with 5, 4, 3, and 2 arguments. Conceptually, it is the same as having the following set of overloaded functions:

```
int min(int a, int b)
{
    return a < b ? a : b;
}

int min(int a, int b, int c)
{
    return min(a, min(b, c));
}

int min(int a, int b, int c, int d)
{
    return min(a, min(b, min(c, d)));
}

int min(int a, int b, int c, int d, int e)
{
    return min(a, min(b, min(c, min(d, e))));
}
```

As a result, min(1, 5, 3, -4, 9) expands to min(1, min(5, min(3, min(-4, 9)))). This can raise questions about the performance of variadic templates. In practice, however, the compilers perform a lot of optimizations, such as inlining as much as possible. The result is that, in practice, when optimizations are enabled, there will be no actual function calls. You can use online resources, such as **Compiler Explorer** (https://godbolt.org/), to see the code generated by different compilers with different options (such as optimization settings). For instance, let's consider the following snippet (where min is the variadic function template with the implementation shown earlier):

```
int main()
{
    std::cout << min(1, 5, 3, -4, 9);
}
```

Compiling this with GCC 11.2 with the -O flag for optimizations produces the following assembly code:

```
sub       rsp, 8
mov       esi, -4
mov       edi, OFFSET FLAT:_ZSt4cout
call      std::basic_ostream<char, std::char_traits<char>>
              ::operator<<(int)
mov       eax, 0
add       rsp, 8
ret
```

You don't need to be an expert in assembly to understand what's happening here. The evaluation of the call to min(1, 5, 3, -4, 9) is done at compile-time and the result, -4, is loaded directly into the ESI register. There are no runtime calls, in this particular case, or computation, since everything is known at compile-time. Of course, that is not necessarily always the case.

The following snippet shows an invocation on the min function template that cannot be evaluated at compile-time because its arguments are only known at runtime:

```
int main()
{
    int a, b, c, d, e;
    std::cin >> a >> b >> c >> d >> e;
    std::cout << min(a, b, c, d, e);
}
```

This time, the assembly code generated is the following (only showing here the code for the call to the min function):

```
mov       esi, DWORD PTR [rsp+12]
mov       eax, DWORD PTR [rsp+16]
cmp       esi, eax
cmovg     esi, eax
mov       eax, DWORD PTR [rsp+20]
cmp       esi, eax
cmovg     esi, eax
mov       eax, DWORD PTR [rsp+24]
cmp       esi, eax
```

```
cmovg   esi, eax
mov     eax, DWORD PTR [rsp+28]
cmp     esi, eax
cmovg   esi, eax
mov     edi, OFFSET FLAT:_ZSt4cout
call    std::basic_ostream<char, std::char_traits<char>>
            ::operator<<(int)
```

We can see from this listing that the compiler has inlined all the calls to the `min` overloads. There is only a series of instructions for loading values into registers, comparisons of register values, and jumps based on the comparison result, but there are no function calls.

When optimizations are disabled, function calls do occur. We can trace these calls that occur during the invocation of the `min` function by using compiler-specific macros. GCC and Clang provide a macro called `__PRETTY_FUNCTION__` that contains the signature of a function and its name. Similarly, Visual C++ provides a macro, called `__FUNCSIG__`, that does the same. These could be used within the body of a function to print its name and signature. We can use them as follows:

```
template <typename T>
T min(T a, T b)
{
#if defined(__clang__) || defined(__GNUC__) || defined(__
GNUG__)
    std::cout << __PRETTY_FUNCTION__ << "\n";
#elif defined(_MSC_VER)
    std::cout << __FUNCSIG__ << "\n";
#endif
    return a < b ? a : b;
}


template <typename T, typename... Args>
T min(T a, Args... args)
{
#if defined(__clang__) || defined(__GNUC__) || defined(__
GNUG__)
    std::cout << __PRETTY_FUNCTION__ << "\n";
#elif defined(_MSC_VER)
```

```
    std::cout << __FUNCSIG__ << "\n";
#endif
    return min(a, min(args...));
}


int main()
{
    min(1, 5, 3, -4, 9);
}
```

The result of the execution of this program, when compiled with Clang, is the following:

```
T min(T, Args...) [T = int, Args = <int, int, int, int>]
T min(T, Args...) [T = int, Args = <int, int, int>]
T min(T, Args...) [T = int, Args = <int, int>]
T min(T, T) [T = int]
T min(T, T) [T = int]
T min(T, T) [T = int]
T min(T, T) [T = int]
```

On the other hand, when compiled with Visual C++, the output is the following:

```
int __cdecl min<int,int,int,int,int>(int,int,int,int,int)
int __cdecl min<int,int,int,int>(int,int,int,int)
int __cdecl min<int,int,int>(int,int,int)
int __cdecl min<int>(int,int)
int __cdecl min<int>(int,int)
int __cdecl min<int>(int,int)
int __cdecl min<int>(int,int)
```

Although the way the signature is formatted is significantly different between Clang/GCC on one hand and VC++ on the other hand, they all show the same: first, an overloaded function with five parameters is called, then one with four parameters, then one with three, and, in the end, there are four calls to the overload with two parameters (which marks the end of the expansion).

Understanding the expansion of parameter packs is key to understanding variadic templates. Therefore, we'll explore this topic in detail in the next section.

Parameter packs

A template or function parameter pack can accept zero, one, or more arguments. The standard does not specify any upper limit for the number of arguments, but in practice, compilers may have some. What the standard does is recommend minimum values for these limits but it does not require any compliance on them. These limits are as follows:

- For a function parameter pack, the maximum number of arguments depends on the limit of arguments for a function call, which is recommended to be at least 256.

- For a template parameter pack, the maximum number of arguments depends on the limit of template parameters, which is recommended to be at least 1,024.

The number of arguments in a parameter pack can be retrieved at compile time with the sizeof... operator. This operator returns a constexpr value of the std::size_t type. Let's see this at work in a couple of examples.

In the first example, the sizeof... operator is used to implement the end of the recursion pattern of the variadic function template sum with the help of a constexpr if statement. If the number of the arguments in the parameter pack is zero (meaning there is a single argument to the function) then we are processing the last argument, so we just return the value. Otherwise, we add the first argument to the sum of the remaining ones. The implementation looks as follows:

```
template <typename T, typename... Args>
T sum(T a, Args... args)
{
    if constexpr (sizeof...(args) == 0)
        return a;
    else
        return a + sum(args...);
}
```

This is semantically equivalent, but on the other hand more concise, than the following classical approach for the variadic function template implementation:

```
template <typename T>
T sum(T a)
{
    return a;
}
```

```
template <typename T, typename... Args>
T sum(T a, Args... args)
{
    return a + sum(args...);
}
```

Notice that `sizeof...(args)` (the function parameter pack) and `sizeof...(Args)` (the template parameter pack) return the same value. On the other hand, `sizeof...(args)` and `sizeof(args)...` are not the same thing. The former is the `sizeof` operator used on the parameter pack `args`. The latter is an expansion of the parameter pack `args` on the `sizeof` operator. These are both shown in the following example:

```
template<typename... Ts>
constexpr auto get_type_sizes()
{
    return std::array<std::size_t,
                      sizeof...(Ts)>{sizeof(Ts)...};
}


auto sizes = get_type_sizes<short, int, long, long long>();
```

In this snippet, `sizeof...(Ts)` evaluates to 4 at compile-time, while `sizeof(Ts)...` is expanded to the following comma-separated pack of arguments: `sizeof(short)`, `sizeof(int)`, `sizeof(long)`, `sizeof(long long)`. Conceptually, the preceding function template, `get_type_sizes`, is equivalent to the following function template with four template parameters:

```
template<typename T1, typename T2,
         typename T3, typename T4>
constexpr auto get_type_sizes()
{
    return std::array<std::size_t, 4> {
       sizeof(T1), sizeof(T2), sizeof(T3), sizeof(T4)
    };
}
```

Typically, the parameter pack is the trailing parameter of a function or template. However, if the compiler can deduce the arguments, then a parameter pack can be followed by other parameters including more parameter packs. Let's consider the following example:

```
template <typename... Ts, typename... Us>
constexpr auto multipacks(Ts... args1, Us... args2)
{
    std::cout << sizeof...(args1) << ','
              << sizeof...(args2) << '\n';
}
```

This function is supposed to take two sets of elements of possibly different types and do something with them. It can be invoked such as in the following examples:

```
multipacks<int>(1, 2, 3, 4, 5, 6);
             // 1,5
multipacks<int, int, int>(1, 2, 3, 4, 5, 6);
             // 3,3
multipacks<int, int, int, int>(1, 2, 3, 4, 5, 6);
             // 4,2
multipacks<int, int, int, int, int, int>(1, 2, 3, 4, 5, 6);
             // 6,0
```

For the first call, the `args1` pack is specified at the function call (as in `multipacks<int>`) and contains 1, and `args2` is deduced to be 2, 3, 4, 5, 6 from the function arguments. Similarly, for the second call, the two packs will have an equal number of arguments, more precisely 1, 2, 3 and 3, 4, 6. For the last call, the first pack contains all the elements, and the second pack is empty. In all these examples, all the elements are of the int type. However, in the following examples, the two packs contain elements of different types:

```
multipacks<int, int>(1, 2, 4.0, 5.0, 6.0);         // 2,3
multipacks<int, int, int>(1, 2, 3, 4.0, 5.0, 6.0); // 3,3
```

For the first call, the `args1` pack will contain the integers 1, 2 and the `args2` pack will be deduced to contain the double values 4.0, 5.0, 6.0. Similarly, for the second call, the `args1` pack will be 1, 2, 3 and the `args2` pack will contain 4.0, 5.0, 6.0.

However, if we change the function template `multipacks` a bit by requiring that the packs be of equal size, then only some of the calls shown earlier would still be possible. This is shown in the following example:

```
template <typename... Ts, typename... Us>
constexpr auto multipacks(Ts... args1, Us... args2)
{
    static_assert(
        sizeof...(args1) == sizeof...(args2),
        "Packs must be of equal sizes.");
}
```

```
multipacks<int>(1, 2, 3, 4, 5, 6);                        // error
multipacks<int, int, int>(1, 2, 3, 4, 5, 6);              // OK
multipacks<int, int, int, int>(1, 2, 3, 4, 5, 6);         // error
multipacks<int, int, int, int, int, int>(1, 2, 3, 4, 5, 6);
                                                          // error
multipacks<int, int>(1, 2, 4.0, 5.0, 6.0);                // error
multipacks<int, int, int>(1, 2, 3, 4.0, 5.0, 6.0);        // OK
```

In this snippet, only the second and the sixth calls are valid. In these two cases, the two deduced packs have three elements each. In all the other cases, as resulting from the prior example, the packs have different sizes and the `static_assert` statement will generate an error at compile-time.

Multiple parameter packs are not specific to variadic function templates. They can also be used for variadic class templates in partial specialization, provided that the compiler can deduce the template arguments. To exemplify this, we'll consider the case of a class template that represents a pair of function pointers. The implementation should allow for storing pointers to any function. To implement this, we define a primary template, called here `func_pair`, and a partial specialization with four template parameters:

- A type template parameter for the return type of the first function
- A template parameter pack for the parameter types of the first function
- A second type template parameter for the return type of the second function
- A second template parameter pack for the parameter types of the second function

The `func_pair` class template is shown in the next listing:

```
template<typename, typename>
struct func_pair;

template<typename R1, typename... A1,
         typename R2, typename... A2>
struct func_pair<R1(A1...), R2(A2...)>
{
    std::function<R1(A1...)> f;
    std::function<R2(A2...)> g;
};
```

To demonstrate the use of this class template, let's also consider the following two functions:

```
bool twice_as(int a, int b)
{
    return a >= b*2;
}


double sum_and_div(int a, int b, double c)
{
    return (a + b) / c;
}
```

We can instantiate the `func_pair` class template and use it to call these two functions as shown in the following snippet:

```
func_pair<bool(int, int), double(int, int, double)> funcs{
    twice_as, sum_and_div };

funcs.f(42, 12);
funcs.g(42, 12, 10.0);
```

Parameter packs can be expanded in a variety of contexts and this will make the topic of the next section.

Understanding parameter packs expansion

Parameter packs can appear in a multitude of contexts. The form of their expansion may depend on this context. These possible contexts are listed ahead along with examples:

- **Template parameter list**: This is for when you specify parameters for a template:

```
template <typename... T>
struct outer
{
    template <T... args>
    struct inner {};
};


outer<int, double, char[5]> a;
```

- **Template argument list**: This is when you specify arguments for a template:

```
template <typename... T>
struct tag {};

template <typename T, typename U, typename ... Args>
void tagger()
{
    tag<T, U, Args...> t1;
    tag<T, Args..., U> t2;
    tag<Args..., T, U> t3;
    tag<U, T, Args...> t4;
}
```

- **Function parameter list**: This is for when you specify parameters for a function template:

```
template <typename... Args>
void make_it(Args... args)
{
}


make_it(42);
make_it(42, 'a');
```

- **Function argument list**: When the expansion pack appears inside the parenthesis of a function call, the largest expression or brace initialization list to the left of the ellipsis is the pattern that is expanded:

```
template <typename T>
T step_it(T value)
{
    return value+1;
}

template <typename... T>
int sum(T... args)
{
    return (... + args);
}

template <typename... T>
void do_sums(T... args)
{
    auto s1 = sum(args...);
    // sum(1, 2, 3, 4)

    auto s2 = sum(42, args...);
    // sum(42, 1, 2, 3, 4)

    auto s3 = sum(step_it(args)...);
    // sum(step_it(1), step_it(2),... step_it(4))
}

do_sums(1, 2, 3, 4);
```

- **Parenthesized initializers**: When the expansion pack appears inside the parenthesis of a direct initializer, function-style cast, member initializer, new expression, and other similar contexts, the rules are the same as for the context of function argument lists:

```
template <typename... T>
struct sum_wrapper
{
```

```
    sum_wrapper(T... args)
    {
        value = (... + args);
    }

    std::common_type_t<T...> value;
};

template <typename... T>
void parenthesized(T... args)
{
    std::array<std::common_type_t<T...>,
                sizeof...(T)> arr {args...};
    // std::array<int, 4> {1, 2, 3, 4}

    sum_wrapper sw1(args...);
    // value = 1 + 2 + 3 + 4

    sum_wrapper sw2(++args...);
    // value = 2 + 3 + 4 + 5
}

parenthesized(1, 2, 3, 4);
```

- **Brace-enclosed initializers**: This is when you perform initialization using the brace notation:

```
template <typename... T>
void brace_enclosed(T... args)
{
    int arr1[sizeof...(args) + 1] = {args..., 0};
    // arr1: {1,2,3,4,0}

    int arr2[sizeof...(args)] = { step_it(args)... };
    // arr2: {2,3,4,5}
}

brace_enclosed(1, 2, 3, 4);
```

• **Base specifiers and member initializer lists**: A pack expansion may specify the list of base classes in a class declaration. In addition, it may also appear in the member initializer list, as this may be necessary to call the constructors of the base classes:

```
struct A {};
struct B {};
struct C {};

template<typename... Bases>
struct X : public Bases...
{
    X(Bases const & ... args) : Bases(args)...
    { }
};

A a;
B b;
C c;
X x(a, b, c);
```

• **Using declarations**: In the context of deriving from a pack of base classes, it may also be useful to be able to introduce names from the base classes into the definition of the derived class. Therefore, a pack expansion may also appear in a using declaration. This is demonstrated based on the previous example:

```
struct A
{
    void execute() { std::cout << "A::execute\n"; }
};

struct B
{
    void execute() { std::cout << "B::execute\n"; }
};

struct C
{
    void execute() { std::cout << "C::execute\n"; }
```

```
};

template<typename... Bases>
struct X : public Bases...
{
    X(Bases const & ... args) : Bases(args)...
    {}

    using Bases::execute...;
};

A a;
B b;
C c;
X x(a, b, c);

x.A::execute();
x.B::execute();
x.C::execute();
```

- **Lambda captures**: The capture clause of a lambda expression may contain a pack expansion, as shown in the following example:

```
template <typename... T>
void captures(T... args)
{
    auto l = [args...]{
                return sum(step_it(args)...); };
    auto s = l();
}

captures(1, 2, 3, 4);
```

- **Fold expressions**: These will be discussed in detail in the following section in this chapter:

```
template <typename... T>
int sum(T... args)
```

```
{
    return (... + args);
}
```

- The sizeof... operator: Examples have already been shown earlier in this section. Here is one again:

```
template <typename... T>
auto make_array(T... args)
{
    return std::array<std::common_type_t<T...>,
                      sizeof...(T)> {args...};
};

auto arr = make_array(1, 2, 3, 4);
```

- **Alignment specifier**: A pack expansion in an alignment specifier has the same effect as having multiple alignas specifiers applied to the same declaration. The parameter pack can be either a type or non-type pack. Examples for both cases are listed here:

```
template <typename... T>
struct alignment1
{
    alignas(T...) char a;
};

template <int... args>
struct alignment2
{
    alignas(args...) char a;
};

alignment1<int, double> al1;
alignment2<1, 4, 8> al2;
```

- **Attribute list**: This is not supported by any compiler yet.

Now that we have learned more about parameter packs and their expansion we can move forward and explore variadic class templates.

Variadic class templates

Class templates may also have a variable number of template arguments. This is key to building some categories of types, such as `tuple` and `variant`, that are available in the standard library. In this section, we will see how we could write a simple implementation for a `tuple` class. A tuple is a type that represents a fixed-size collection of heterogeneous values.

When implementing variadic function templates we used a recursion pattern with two overloads, one for the general case and one for ending the recursion. The same approach has to be taken with variadic class templates, except that we need to use specialization for this purpose. Next, you can see a minimal implementation for a tuple:

```
template <typename T, typename... Ts>
struct tuple
{
    tuple(T const& t, Ts const &... ts)
        : value(t), rest(ts...)
    {
    }

    constexpr int size() const { return 1 + rest.size(); }

    T              value;
    tuple<Ts...> rest;
};

template <typename T>
struct tuple<T>
{
    tuple(const T& t)
        : value(t)
    {
    }

    constexpr int size() const { return 1; }

    T value;
};
```

The first class is the primary template. It has two template parameters: a type template and a parameter pack. This means, at the minimum, there must be one type specified for instantiating this template. The primary template tuple has two member variables: `value`, of the `T` type, and `rest`, of type `tuple<Ts...>`. This is an expansion of the rest of the template arguments. This means a tuple of N elements will contain the first element and another tuple; this second tuple, in turn, contains the second element and yet another tuple; this third nested tuple contains the rest. And this pattern continues until we end up with a tuple with a single element. This is defined by the partial specialization `tuple<T>`. Unlike the primary template, this specialization does not aggregate another tuple object.

We can use this simple implementation to write code like the following:

```
tuple<int> one(42);
tuple<int, double> two(42, 42.0);
tuple<int, double, char> three(42, 42.0, 'a');

std::cout << one.value << '\n';
std::cout << two.value << ','
          << two.rest.value << '\n';
std::cout << three.value << ','
          << three.rest.value << ','
          << three.rest.rest.value << '\n';
```

Although this works, accessing elements through the `rest` member, such as in `three.rest.rest.value`, is very cumbersome. And the more elements a tuple has the more difficult it is to write code in this way. Therefore, we'd like to use some helper function to simplify accessing the elements of a tuple. The following is a snippet of how the previous could be transformed:

```
std::cout << get<0>(one) << '\n';
std::cout << get<0>(two) << ','
          << get<1>(two) << '\n';
std::cout << get<0>(three) << ','
          << get<1>(three) << ','
          << get<2>(three) << '\n';
```

Here, get<N> is a variadic function template that takes a tuple as an argument and returns a reference to the element at the N index in the tuple. Its prototype could look like the following:

```
template <size_t N, typename... Ts>
typename nth_type<N, Ts...>::value_type & get(tuple<Ts...>& t);
```

The template arguments are the index and a parameter pack of the tuple types. Its implementation, however, requires some helper types. First, we need to know what the type of the element is at the N index in the tuple. This can be retrieved with the help of the following nth_type variadic class template:

```
template <size_t N, typename T, typename... Ts>
struct nth_type : nth_type<N - 1, Ts...>
{
    static_assert(N < sizeof...(Ts) + 1,
                  "index out of bounds");
};

template <typename T, typename... Ts>
struct nth_type<0, T, Ts...>
{
    using value_type = T;
};
```

Again, we have a primary template that uses recursive inheritance, and the specialization for the index 0. The specialization defines an alias called value_type for the first type template (which is the head of the list of template arguments). This type is only used as a mechanism for determining the type of a tuple element. We need another variadic class template for retrieving the value. This is shown in the following listing:

```
template <size_t N>
struct getter
{
    template <typename... Ts>
    static typename nth_type<N, Ts...>::value_type&
    get(tuple<Ts...>& t)
    {
        return getter<N - 1>::get(t.rest);
```

```
    }
};


template <>
struct getter<0>
{
    template <typename T, typename... Ts>
    static T& get(tuple<T, Ts...>& t)
    {
        return t.value;
    }
};
```

We can see here the same recursive pattern, with a primary template and an explicit specialization. The class template is called getter and has a single template parameter, which is a non-type template parameter. This represents the index of the tuple element we want to access. This class template has a static member function called get. This is a variadic function template. The implementation in the primary template calls the get function with the rest member of the tuple as an argument. On the other hand, the implementation of the explicit specialization returns the reference to the member value of the tuple.

With all these defined, we can now provide an actual implementation for the helper variadic function template get. This implementation relies on the getter class template and calls its get variadic function template:

```
template <size_t N, typename... Ts>
typename nth_type<N, Ts...>::value_type &
get(tuple<Ts...>& t)
{
    return getter<N>::get(t);
}
```

If this example seems a little bit complicated, perhaps analyzing it step by step will help you better understand how it all works. Therefore, let's start with the following snippet:

```
tuple<int, double, char> three(42, 42.0, 'a');
get<2>(three);
```

We will use the cppinsights.io web tools to check the template instantiations that occur from this snippet. The first to look at is the class template tuple. We have a primary template and several specializations, as follows:

```
template <typename T, typename... Ts>
struct tuple
{
    tuple(T const& t, Ts const &... ts)
        : value(t), rest(ts...)
    { }

    constexpr int size() const { return 1 + rest.size(); }

    T value;
    tuple<Ts...> rest;
};

template<> struct tuple<int, double, char>
{
  inline tuple(const int & t,
              const double & __ts1, const char & __ts2)
    : value{t}, rest{tuple<double, char>(__ts1, __ts2)}
    {}

  inline constexpr int size() const;

  int value;
  tuple<double, char> rest;
};

template<> struct tuple<double, char>
{
  inline tuple(const double & t, const char & __ts1)
    : value{t}, rest{tuple<char>(__ts1)}
    {}
```

```
  inline constexpr int size() const;

  double value;
  tuple<char> rest;
};

template<> struct tuple<char>
{
  inline tuple(const char & t)
  : value{t}
  {}

  inline constexpr int size() const;

  char value;
};

template<typename T>
struct tuple<T>
{
   inline tuple(const T & t) : value{t}
   { }

   inline constexpr int size() const
   { return 1; }

   T value;
};
```

The `tuple<int, double, char>` structure contains an `int` and a
`tuple<double, char>`, which contains a `double` and a `tuple<char>`, which, in
turn, contains a `char` value. This last class represents the end of the recursive definition
of the tuple. This can be conceptually represented graphically as follows:

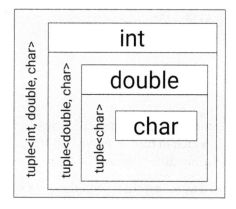

Figure 3.1 – An example tuple

Next, we have the nth_type class template, for which, again, we have a primary template and several specializations, as follows:

```
template <size_t N, typename T, typename... Ts>
struct nth_type : nth_type<N - 1, Ts...>
{
    static_assert(N < sizeof...(Ts) + 1,
                  "index out of bounds");
};

template<>
struct nth_type<2, int, double, char> :
    public nth_type<1, double, char>
{ };

template<>
struct nth_type<1, double, char> : public nth_type<0, char>
{ };

template<>
struct nth_type<0, char>
{
    using value_type = char;
};
```

```
template<typename T, typename ... Ts>
struct nth_type<0, T, Ts...>
{
    using value_type = T;
};
```

The nth_type<2, int, double, char> specialization is derived from nth_type<1, double, char>, which in turn is derived from nth_type<0, char>, which is the last base class in the hierarchy (the end of the recursive hierarchy).

The nth_type structure is used as the return type in the getter helper class template, which is instantiated as follows:

```
template <size_t N>
struct getter
{
    template <typename... Ts>
    static typename nth_type<N, Ts...>::value_type&
    get(tuple<Ts...>& t)
    {
        return getter<N - 1>::get(t.rest);
    }
};

template<>
struct getter<2>
{
    template<>
    static inline typename
    nth_type<2UL, int, double, char>::value_type &
    get<int, double, char>(tuple<int, double,  char> & t)
    {
        return getter<1>::get(t.rest);
    }
};

template<>
struct getter<1>
```

```
{
    template<>
    static inline typename nth_type<1UL, double,
                                    char>::value_type &
    get<double, char>(tuple<double, char> & t)
    {
        return getter<0>::get(t.rest);
    }
};

template<>
struct getter<0>
{
    template<typename T, typename ... Ts>
    static inline T & get(tuple<T, Ts...> & t)
    {
        return t.value;
    }

    template<>
    static inline char & get<char>(tuple<char> & t)
    {
        return t.value;
    }
};
```

Lastly, the get function template that we use to retrieve the value of an element of a tuple is defined as follows:

```
template <size_t N, typename... Ts>
typename nth_type<N, Ts...>::value_type &
get(tuple<Ts...>& t)
{
    return getter<N>::get(t);
}
```

```
template<>
typename nth_type<2UL, int, double, char>::value_type &
get<2, int, double, char>(tuple<int, double, char> & t)
{
    return getter<2>::get(t);
}
```

Should there be more calls to the get function more specializations of get would exist. For instance, for get<1>(three), the following specialization would be added:

```
template<>
typename nth_type<1UL, int, double, char>::value_type &
get<1, int, double, char>(tuple<int, double, char> & t)
{
    return getter<1>::get(t);
}
```

This example helped us demonstrate how to implement variadic class templates with a primary template for the general case and a specialization for the end case of the variadic recursion.

You have probably noticed the use of the keyword typename to prefix the nth_type<N, Ts...>::value_type type, which is a **dependent type**. In C++20, this is no longer necessary. However, this topic will be addressed in detail in *Chapter 4, Advanced Template Concepts*.

Because implementing variadic templates is often verbose and can be cumbersome, the C++17 standard added **fold expressions** to ease this task. We will explore this topic in the next section.

Fold expressions

A **fold expression** is an expression involving a parameter pack that folds (or reduces) the elements of the parameter pack over a binary operator. To understand how this works, we will look at several examples. Earlier in this chapter, we implemented a variable function template called sum that returned the sum of all its supplied arguments. For convenience, we will show it again here:

```
template <typename T>
T sum(T a)
{
```

```
        return a;
}

template <typename T, typename... Args>
T sum(T a, Args... args)
{
        return a + sum(args...);
}
```

With fold expressions, this implementation that requires two overloads can be reduced to the following form:

```
template <typename... T>
int sum(T... args)
{
        return (... + args);
}
```

There is no need for overloaded functions anymore. The expression (... + args) represents the fold expression, which upon evaluation becomes ((((arg0 + arg1) + arg2) + ...) + argN). The enclosing parentheses are part of the fold expression. We can use this new implementation, just as we would use the initial one, as follows:

```
int main()
{
        std::cout << sum(1) << '\n';
        std::cout << sum(1,2) << '\n';
        std::cout << sum(1,2,3,4,5) << '\n';
}
```

There are four different types of folds, which are listed as follows:

Fold	Syntax	Expansion
Unary right fold	(pack op ...)	(arg1 op (... op (argN-1 op argN)))
Unary left fold	(... op pack)	(((arg1 op arg2) op ...) op argN)
Binary right fold	(pack op ... op init)	(arg1 op (... op (argN-1 op (argN op init))))
Binary left fold	(init op ... op pack)	((((init op arg1) op arg2) op ...) op argN)

Table 3.1

In this table, the following names are used:

- `pack` is an expression that contains an unexpanded parameter pack, and `arg1`, `arg2`, `argN-1`, and `argN` are the arguments contained in this pack.
- `op` is one of the following binary operators: `+ - * / % ^ & | = < > << >>` `+= -= *= /= %= ^= &= |= <<= >>= == != <= >= && || , .* ->*`.
- `init` is an expression that does not contain an unexpanded parameter pack.

In a unary fold, if the pack does not contain any elements, only some operators are allowed. These are listed in the following table, along with the value of the empty pack:

Operator	Value of the empty pack
`&&` (logical AND)	`true`
`||` (logical OR)	`false`
`,` (comma operator)	`void()`

Table 3.2

Unary and binary folds differ in the use of an initialization value, that is present only for binary folds. Binary folds have the binary operator repeated twice (it must be the same operator). We can transform the variadic function template `sum` from using a unary right fold expression into one using a binary right fold by including an initialization value. Here is an example:

```
template <typename... T>
int sum_from_zero(T... args)
{
    return (0 + ... + args);
}
```

One could say there is no difference between the `sum` and `sum_from_zero` function templates. That is not actually true. Let's consider the following invocations:

```
int s1 = sum();           // error
int s2 = sum_from_zero(); // OK
```

Calling `sum` without arguments will produce a compiler error, because unary fold expressions (over the operator + in this case) must have non-empty expansions. However, binary fold expressions do not have this problem, so calling `sum_from_zero` without arguments works and the function will return `0`.

In these two examples with sum and sum_from_zero, the parameter pack args appears directly within the fold expression. However, it can be part of an expression, as long as it is not expanded. This is shown in the following example:

```
template <typename... T>
void printl(T... args)
{
    (..., (std::cout << args)) << '\n';
}


template <typename... T>
void printr(T... args)
{
    ((std::cout << args), ...) << '\n';
}
```

Here, the parameter pack args is part of the (std::cout << args) expression. This is not a fold expression. A fold expression is ((std::cout << args), ...). This is a unary left fold over the comma operator. The printl and printr functions can be used as in the following snippet:

```
printl('d', 'o', 'g');   // dog
printr('d', 'o', 'g');   // dog
```

In both these cases, the text printed to the console is dog. This is because the unary left fold expands to (((std::cout << 'd'), std::cout << 'o'), << std::cout << 'g') and the unary right fold expands to (std::cout << 'd', (std::cout << 'o', (std::cout << 'g'))) and these two are evaluated in the same way. This is because a pair of expressions separated by a comma is evaluated left to right. This is true for the built-in comma operator. For types that overload the comma operator, the behavior depends on how the operator is overloaded. However, there are very few corner cases for overloading the comma operator (such as simplifying indexing multi-dimensional arrays). Libraries such as Boost.Assign and SOCI overload the comma operator, but, in general, this is an operator you should avoid overloading.

Let's consider another example for using the parameter pack in an expression inside a fold expression. The following variadic function template inserts multiple values to the end of a std::vector:

```
template<typename T, typename... Args>
void push_back_many(std::vector<T>& v, Args&&... args)
```

```
{
    (v.push_back(args), ...);
}

push_back_many(v, 1, 2, 3, 4, 5); // v = {1, 2, 3, 4, 5}
```

The parameter pack `args` is used with the `v.push_back(args)` expression that is folded over the comma operator. The unary left fold expression is `(v.push_back(args), ...)`.

Fold expressions have several benefits over the use of recursion to implement variadic templates. These benefits are as follows:

- Less and simpler code to write.

- Fewer template instantiations, which leads to faster compile times.

- Potentially faster code since multiple function calls are replaced with a single expression. However, this point may not be true in practice, at least not when optimizations are enabled. We have already seen that the compilers optimize code by removing these function calls.

Now that we have seen how to create variadic function templates, variadic class templates, and how to use fold expressions, we are left to discuss the other kinds of templates that can be variadic: alias templates and variable templates. We will start with the former.

Variadic alias templates

Everything that can be templatized can also be made variadic. An alias template is an alias (another name) for a family of types. A variadic alias template is a name for a family of types with a variable number of template parameters. With the knowledge accumulated so far, it should be fairly trivial to write alias templates. Let's see an example:

```
template <typename T, typename... Args>
struct foo
{
};

template <typename... Args>
using int_foo = foo<int, Args...>;
```

The class template `foo` is variadic and takes at least one type template argument. `int_foo`, on the other hand, is only a different name for a family of types instantiated from the `foo` type with `int` as the first type template arguments. These could be used as follows:

```
foo<double, char, int> f1;
foo<int, char, double> f2;
int_foo<char, double> f3;
static_assert(std::is_same_v<decltype(f2), decltype(f3)>);
```

In this snippet, `f1` on one hand and `f2` and `f3` on the other are instances of different `foo` types, as they are instantiated from different sets of template arguments for `foo`. However, `f2` and `f3` are instances of the same type, `foo<int, char, double>`, since `int_foo<char, double>` is just an alias for this type.

A similar example, although a bit more complex, is presented ahead. The standard library contains a class template called `std::integer_sequence`, which represents a compile-time sequence of integers, along with a bunch of alias templates to help create various kinds of such integer sequences. Although the code shown here is a simplified snippet, their implementation can, at least conceptually, be as follows:

```
template<typename T, T... Ints>
struct integer_sequence
{};

template<std::size_t... Ints>
using index_sequence = integer_sequence<std::size_t,
                                        Ints...>;

template<typename T, std::size_t N, T... Is>
struct make_integer_sequence :
  make_integer_sequence<T, N - 1, N - 1, Is...>
{};

template<typename T, T... Is>
struct make_integer_sequence<T, 0, Is...> :
  integer_sequence<T, Is...>
{};
```

```
template<std::size_t N>
using make_index_sequence = make_integer_sequence<std::size_t,
                                                    N>;


template<typename... T>
using index_sequence_for =
    make_index_sequence<sizeof...(T)>;
```

There are three alias templates here:

- `index_sequence`, which creates an `integer_sequence` for the `size_t` type; this is a variadic alias template.

- `index_sequence_for`, which creates an `integer_sequence` from a parameter pack; this is also a variadic alias template.

- `make_index_sequence`, which creates an `integer_sequence` for the `size_t` type with the values 0, 1, 2, …, *N-1*. Unlike the previous ones, this is not an alias for a variadic template.

The last subject to address in this chapter is variadic variable templates.

Variadic variable templates

As mentioned before, variable templates may also be variadic. However, variables cannot be defined recursively, nor can they be specialized like class templates. Fold expressions, which simplify generating expressions from a variable number of arguments, are very handy for creating variadic variable templates.

In the following example, we define a variadic variable template called Sum that is initialized at compile-time with the sum of all integers supplied as non-type template arguments:

```
template <int... R>
constexpr int Sum = (... + R);


int main()
{
    std::cout << Sum<1> << '\n';
    std::cout << Sum<1,2> << '\n';
```

```
        std::cout << Sum<1,2,3,4,5> << '\n';
}
```

This is similar to the sum function written with the help of fold expressions. However, in that case, the numbers to add were provided as function arguments. Here, they are provided as template arguments to the variable template. The difference is mostly syntactic; with optimizations enabled, the end result is likely the same in terms of generated assembly code, and therefore performance.

Variadic variable templates follow the same patterns as all the other kinds of templates although they are not used as much as the others. However, by concluding this topic we have now completed the learning of variadic templates in C++.

Summary

In this chapter, we have explored an important category of templates, variadic templates, which are templates with a variable number of template arguments. We can create variadic function templates, class templates, variable templates, and alias templates. The techniques to create variadic function templates and variadic class templates are different but incur a form of compile-time recursion. For the latter, this is done with template specialization, while for the former with function overloads. Fold expressions help to expand a variable number of arguments into a single expression, avoiding the need of using function overloads and enabling the creation of some categories of variadic variable templates such as the ones we have previously seen.

In the next chapter, we will look into a series of more advanced features that will help you consolidate your knowledge of templates.

Questions

1. What are variadic templates and why are they useful?
2. What is a parameter pack?
3. What are the contexts where parameter packs can be expanded?
4. What are fold expressions?
5. What are the benefits of using fold expressions?

Further reading

- *C++ Fold Expressions 101*, Jonathan Boccara, `https://www.fluentcpp.com/2021/03/12/cpp-fold-expressions/`

- *Fold Expressions in C++ 17*, Vaibhav, `https://mainfunda.com/fold-expressions-in-cpp17/`

- *Nifty Fold Expression Tricks*, Jonathan Müller, `https://www.foonathan.net/2020/05/fold-tricks/`

Part 2:
Advanced
Template Features

In this part, you will explore a variety of advanced features, including name binding and dependent names, template recursion, template argument deduction, and forwarding references. Here, you will learn about type traits that help us query information about types and perform conditional compilation using various language features. Furthermore, you will learn how to specify requirements on template arguments with C++20 concepts and constraints and explore the content of the standard concepts library.

This part comprises the following chapters:

- *Chapter 4, Advanced Template Concepts*
- *Chapter 5, Type Traits and Conditional Compilation*
- *Chapter 6, Concepts and Constraints*

4

Advanced Template Concepts

In the previous chapters, we learned the core fundamentals of C++ templates. At this point, you should be able to write templates that are perhaps not very complex. However, there are many more details concerning templates, and this chapter is dedicated to these more advanced topics. These include the following topics that we address next:

- Understanding name binding and dependent names
- Exploring template recursion
- Understanding template argument deduction
- Learning forwarding references and perfect forwarding
- Using the `decltype` specifier and the `std::declval` type operator
- Understanding friendship in templates

On completing this chapter, you will acquire a deeper knowledge of these advanced template concepts and be able to understand and write more complex template code.

We will start this chapter by learning about name binding and dependent names.

Understanding name binding and dependent names

The term **name binding** refers to the process of finding the declaration of each name that is used within a template. There are two kinds of names used within a template: **dependent names** and **non-dependent names**. The former are names that depend on the type or value of a template parameter that can be a type, non-type, or template parameter. Names that don't depend on template parameters are called *non-dependent*. The name lookup is performed differently for dependent and non-dependent names:

- For dependent names, it is performed at *the point of template instantiation*.
- For non-dependent names, it is performed at *the point of the template definition*.

We will first look at non-dependent names. As previously mentioned, name lookup happens at the point of the template definition. This is located immediately before the template definition. To understand how this works, let's consider the following example:

```cpp
template <typename T>
struct processor;          // [1] template declaration
void handle(double value)  // [2] handle(double) definition
{
    std::cout << "processing a double: " << value << '\n';
}


template <typename T>
struct parser               // [3] template definition
{
    void parse()
    {
        handle(42);         // [4] non-dependent name
    }
};


void handle(int value)      // [5] handle(int) definition
{
    std::cout << "processing an int: " << value << '\n';
}


int main()
```

```
{
    parser<int> p;          // [6] template instantiation
    p.parse();
}
```

There are several points of reference that are marked in the comments on the right side. At point [1], we have the declaration of a class template called parser. This is followed at point [2] by the definition of a function called handle that takes a double as its argument. The definition of the class template follows at point [3]. This class contains a single method called run that invokes a function called handle with the value 42 as its argument, at point [4].

The name handle is a non-dependent name because it does not depend on any template parameter. Therefore, name lookup and binding are performed at this point. handle must be a function known at point [3] and the function defined at [2] is the only match. After the class template definition, at point [5] we have the definition of an overload for the function handle, which takes an integer as its argument. This is a better match for handle(42), but it comes after the name binding has been performed, and therefore it will be ignored. In the main function, at point [6], we have an instantiation of the parser class template for the type int. Upon calling the run function, the text processing a double: 42 will be printed to the console output.

The next example is designed to introduce you to the concept of dependent names. Let's look at the code first:

```
template <typename T>
struct handler              // [1] template definition
{
    void handle(T value)
    {
        std::cout << "handler<T>: " << value << '\n';
    }
};


template <typename T>
struct parser               // [2] template definition
{
    void parse(T arg)
    {
        arg.handle(42);     // [3] dependent name
```

```
    }
};


template <>
struct handler<int>      // [4] template specialization
{
    void handle(int value)
    {
        std::cout << "handler<int>: " << value << '\n';
    }
};


int main()
{
    handler<int> h;              // [5] template instantiation
    parser<handler<int>> p;  // [6] template instantiation
    p.parse(h);
}
```

This example is slightly different from the previous one. The `parser` class template is very similar, but the `handle` functions have become members of another class template. Let's analyze it point by point.

At the point mark with `[1]` in the comments, we have the definition of a class template called `handler`. This contains a single, public method called `handle` that takes an argument of the `T` type and prints its value to the console. Next, at point `[2]`, we have the definition of the class template called `parser`. This is similar to the previous one, except for one key aspect: at point `[3]`, it invokes a method called `handle` on its argument. Because the type of the argument is the template parameter `T`, it makes `handle` a dependent name. Dependent names are looked up at the point of template instantiation, so `handle` is not bound at this point. Continuing with the code, at point `[4]`, there is a template specialization for the `handler` class template for the type `int`. As a specialization, this is a better match for the dependent name. Therefore, when the template instantiation happens at point `[6]`, `handler<int>::handle` is the name that is bound to the dependent name used at `[3]`. Running this program will print `handler<int>: 42` to the console.

Now that we've seen how name binding occurs, let's learn how this relates to template instantiation.

Two-phase name lookup

The key takeaway from the previous section is that name lookup happens differently for dependent names (those that depend on a template parameter) and non-dependent names (those that do not depend on a template parameter, plus the template name and names defined in the current template instantiation). When the compiler passes through the definition of a template it needs to figure out whether a name is dependent or non-dependent. Further name lookup depends on this categorization and happens either at the template definition point (for non-dependent names) or the template instantiation point (for dependent names). Therefore, instantiation of a template happens in two phases:

- The first phase occurs at the point of the definition when the template syntax is checked and names are categorized as dependent or non-dependent.

- The second phase occurs at the point of instantiation when the template arguments are substituted for the template parameters. Name binding for dependent names happens at this point.

This process in two steps is called **two-phase name lookup**. To understand it better, let's consider another example:

```cpp
template <typename T>
struct base_parser
{
   void init()
   {
      std::cout << "init\n";
   }
};

template <typename T>
struct parser : base_parser<T>
{
   void parse()
   {
      init();          // error: identifier not found
      std::cout << "parse\n";
   }
};
```

```
int main()
{
    parser<int> p;
    p.parse();
}
```

In this snippet, we have two class templates: `base_parser`, which contains a public method called `init`, and `parser`, which derives from `base_parser` and contains a public method called `parse`. The `parse` member function calls a function called `init` and the intention is that it's the base-class method `init` that is invoked here. However, the compiler will issue an error, because it's not able to find `init`. The reason this happens is that `init` is a non-dependent name (as it does not depend on a template parameter). Therefore, it must be known at the point of the definition of the `parser` template. Although a `base_parser<T>::init` exists, the compiler cannot assume it's what we want to call because the primary template `base_parser` can be later specialized and `init` can be defined as something else (such as a type, or a variable, or another function, or it may be missing entirely). Therefore, name lookup does not happen in the base class, only on its enclosing scope, and there is no function called `init` in `parser`.

This problem can be fixed by making `init` a dependent name. This can be done either by prefixing with `this->` or with `base_parser<T>::`. By turning `init` into a dependent name, its name binding is moved from the point of template definition to the point of template instantiation. In the following snippet, this problem is solved by invoking `init` through the `this` pointer:

```
template <typename T>
struct parser : base_parser<T>
{
    void parse()
    {
        this->init();          // OK

        std::cout << "parse\n";
    }
};
```

Continuing this example, let's consider that a specialization of `base_parser` for the `int` type is made available after the definition of the `parser` class template. This can look as follows:

```
template <>
struct base_parser<int>
{
    void init()
    {
        std::cout << "specialized init\n";
    }
};
```

Furthermore, let's consider the following use of the `parser` class template:

```
int main()
{
    parser<int> p1;
    p1.parse();

    parser<double> p2;
    p2.parse();
}
```

When you run this program, the following text will be printed to the console:

```
specialized init
parse
init
parse
```

The reason for this behavior is that p1 is an instance of `parser<int>` and there is a specialization of its base class, `base_parser<int>` that implements the `init` function and prints `specialized init` to the console. On the other hand, p2 is an instance of `parser<double>`. Since a specialization of `base_parser` for the `double` type is not available, the `init` function from the primary template is being called and this only prints `init` to the console.

The next subject of this broader topic is using dependent names that are types. Let's learn how that works.

Dependent type names

In the examples seen so far, the dependent name was a function or a member function. However, there are cases when a dependent name is a type. This is demonstrated with the following example:

```
template <typename T>
struct base_parser
{
    using value_type = T;
};


template <typename T>
struct parser : base_parser<T>
{
    void parse()
    {
        value_type v{};                    // [1] error
        // or
        base_parser<T>::value_type v{};    // [2] error

        std::cout << "parse\n";
    }
};
```

In this snippet, base_parser is a class template that defines a type alias for T called value_type. The parser class template, which derives from base_parser, needs to use this type within its parse method. However, both value_type and base_parser<T>::value_type do not work, and the compiler is issuing an error. value_type does not work because it's a non-dependent name and therefore it will not be looked up in the base class, only in the enclosing scope. base_parser<T>::value_type does not work either because the compiler cannot assume this is actually a type. A specialization of base_parser may follow and value_type could be defined as something else than a type.

In order to fix this problem, we need to tell the compiler the name refers to a type. Otherwise, by default, the compiler assumes it's not a type. This is done with the `typename` keyword, at the point of definition, shown as follows:

```
template <typename T>
struct parser : base_parser<T>
{
    void parse()
    {
        typename base_parser<T>::value_type v{}; // [3] OK

        std::cout << "parse\n";
    }
};
```

There are actually two exceptions to this rule:

- When specifying a base class
- When initializing class members

Let's see an example for these two exceptions:

```
struct dictionary_traits
{
    using key_type = int;
    using map_type = std::map<key_type, std::string>;
    static constexpr int identity = 1;
};


template <typename T>
struct dictionary : T::map_type          // [1]
{
    int start_key { T::identity };    // [2]
    typename T::key_type next_key;    // [3]
};


int main()
{
```

```
      dictionary<dictionary_traits> d;
}
```

The `dictionay_traits` is a class used as the template argument for the `dictionary` class template. This class derives from `T::map_type` (see line [1]) but the use of the `typename` keyword is not required here. The dictionary class defines a member called `start_key`, which is an `int` initialized with the value of `T::identity` (see line [2]). Again, the `typename` keyword is not needed here. However, if we want to define yet another member of the type `T::key_type` (see line [3]) we do need to use `typename`.

The requirements for using `typename` have been relaxed in **C++20** making the use of type names easier. The compiler is now able to deduce that we are referring to a type name in a multitude of contexts. For instance, defining a member variable as we did on line [3] previously no longer requires prefixing with the `typename` keyword.

In C++20, `typename` is implicit (can be deduced by the compiler) in the following contexts:

- In using declarations

- In the declaration of data members

- In the declaration or definition of function parameters

- In trailing return types

- In default arguments of type-parameters of a template

- In the type-id of a `static_cast`, `const_cast`, `reinterpret_cast`, or `dynamic_cast` statement

Some of these contexts are exemplified in the following snippet:

```
template <typename T>
struct dictionary : T::map_type
{
    int start_key{ T::identity };
    T::key_type next_key;                        // [1]

    using value_type = T::map_type::mapped_type; // [2]

    void add(T::key_type const&, value_type const&) {} // [3]
};.
```

At all the lines marked with `[1]`, `[2]`, and `[3]` in this snippet, prior to C++20, the `typename` keyword was required to indicate a type name (such as `T::key_type` or `T::map_type::mapped_type`). When compiled with C++20, this is no longer necessary.

> **Note**
>
> In *Chapter 2, Template Fundamentals*, we have seen that the keywords `typename` and `class` can be used to introduce type template parameters and they are interchangeable. The keyword `typename` here, although it has a similar purpose, cannot be substituted with the keyword `class`.

Not only types can be dependent names but other templates too. We look at this topic in the next subsection.

Dependent template names

There are cases when the dependent name is a template, such as a function template or a class template. However, the default behavior of the compiler is to interpret the dependent name as a non-type, which leads to errors concerning the usage of the comparison operator <. Let's demonstrate this with an example:

```
template <typename T>
struct base_parser
{
    template <typename U>
    void init()
    {
        std::cout << "init\n";
    }
};


template <typename T>
struct parser : base_parser<T>
{
    void parse()
    {
```

```
        //  base_parser<T>::init<int>();           // [1] error
        base_parser<T>::template init<int>();   // [2] OK


        std::cout << "parse\n";
    }
};
```

This is similar to the previous snippets, but the init function in base_parser is also a template. The attempt to call it using the base_parser<T>::init<int>() syntax, as seen at point [1], results in a compiler error. Therefore, we must use the template keyword to tell the compiler the dependent name is a template. This is done as shown at point [2].

Keep in mind that the template keyword can only follow the scope resolution operator (::), member access through pointer (->), and the member access (.). Examples of correct usage are X::template foo<T>(), this->template foo<T>(), and obj.template foo<T>().

The dependent name does not have to be a function template. It can also be a class template, shown as follows:

```
template <typename T>
struct base_parser
{
    template <typename U>
    struct token {};
};


template <typename T>
struct parser : base_parser<T>
{
    void parse()
    {
```

```
        using token_type =
            base_parser<T>::template token<int>; // [1]
        token_type t1{};

        typename base_parser<T>::template token<int> t2{};
                                                    // [2]

        std::cout << "parse\n";
    }
};
```

The token class is an inner class template of the base_parser class template. It can be either used as in the line marked with [1], where a type alias is defined (which is then used to instantiate an object) or as at line [2], where it is used directly to declare a variable. Notice that the typename keyword is not necessary at [1], where the using declaration indicates we are dealing with a type, but is required at [2] because the compiler would otherwise assume it's a non-type name.

The use of the typename and template keywords is not required in some contexts of the current template instantiation being observed. This will be the topic of the next subsection.

Current instantiation

The requirement to use the typename and template keywords to disambiguate dependent names may be avoided in the context of a class template definition where the compiler is able to deduce some dependent names (such as the name of a nested class) to refer to the current instantiation. This means some errors can be identified sooner, at the point of definition instead of the point of instantiation.

The complete list of names that can refer to the current instantiation, according to the C++ *Standard, §13.8.2.1 - Dependent Types*, is presented in the following table:

Context	Names
Class template definition	Nested class
	Member of class template
	Member of a nested class
	Injected class-name of the template
	Injected class-name of a nested class
Primary class template definition or Definition of members of a primary class template	Name of the class template followed by a template argument list for the primary template where each argument is equivalent to its corresponding parameter
Definition of a nested class or class template	Name of a nested class used as a member of the current instantiation
Definition of a partial specialization or Definition of a member of a partial specialization	Name of the class template followed by a template argument list for the partial specialization, where each argument is equivalent to its corresponding parameter

Table 4.1

The following are the rules for considering a name as part of the current instantiation:

- An unqualified name (that does not appear on the right side of the scope resolution operator ::) found in the current instantiation or its non-dependent base

- A qualified name (that appears on the right side of the scope resolution operator ::) if its qualifier (the part that appears on the left side of the scope resolution operator) names the current instantiation and is found in the current instantiation or its non-dependent base

- A name used in a class member access expression where the object expression is the current instantiation and the name is found in the current instantiation or its non-dependent base

> **Note**
>
> It is said that a base class is a **dependent class** if it is a dependent type (depends on a template parameter) and is not in the current instantiation. Otherwise, a base class is said to be a **non-dependent class**.

These rules may sound a bit harder to comprehend; therefore, let's try to understand them with the help of several examples, as follows:

```cpp
template <typename T>
struct parser
{
    parser* p1;            // parser is the CI
    parser<T>* p2;         // parser<T> is the CI
    ::parser<T>* p3;       // ::parser<T> is the CI
    parser<T*> p4;         // parser<T*> is not the CI

    struct token
    {
        token* t1;                      // token is the CI
        parser<T>::token* t2;           // parser<T>::token is the CI
        typename parser<T*>::token* t3;
                            // parser<T*>::token is not the CI
    };
};

template <typename T>
struct parser<T*>
{
    parser<T*>* p1;    // parser<T*> is the CI
    parser<T>* p2;     // parser<T> is not the CI
};
```

In the primary template `parser`, the names `parser`, `parser<T>`, and `::parser<T>` all refer to the current instantiation. However, `parser<T*>` does not. The class `token` is a nested class of the primary template `parser`. In the scope of this class, `token` and `parser<T>::token` are both denoting the current instantiation. The same is not true for `parser<T*>::token`. This snippet also contains a partial specialization of the primary template for the pointer type `T*`. In the context of this partial specialization, `parser<T*>` is the current instantiation, but `parser<T>` is not.

Dependent names are an important aspect of template programming. The key takeaway from this section is that names are categorized as dependent (those that depend on a template parameter) and non-dependent (those that don't depend on a template parameter). Name binding happens at the point of definition for non-dependent types and at the point of instantiation for dependent types. In some cases, the keywords `typename` and `template` are required to disambiguate the use of names and tell the compiler that a name refers to a type or a template. In the context of a class template definition, the compiler is, however, able to figure out that some dependent names refer to the current instantiation, which enables it to identify errors sooner.

In the next section, we move our attention to a topic that we briefly touched already, which is template recursion.

Exploring template recursion

In *Chapter 3*, *Variadic Templates*, we discussed variadic templates and saw that they are implemented with a mechanism that looks like recursion. In fact, it is overloaded functions and class template specializations respectively. However, it is possible to create recursive templates. To demonstrate how this works, we'll look at implementing a compile-time version of the factorial function. This is typically implemented in a recursive manner, and a possible implementation is the following:

```
constexpr unsigned int factorial(unsigned int const n)
{
    return n > 1 ? n * factorial(n - 1) : 1;
}
```

This should be trivial to understand: return the result of multiplying the function argument with the value returned by calling the function recursively with the decremented argument, or return the value 1 if the argument is 0 or 1. The type of the argument (and the return value) is `unsigned int` to avoid calling it for negative integers.

To compute the value of the factorial function at compile time, we need to define a class template that contains a data member holding the value of the function. The implementation looks as follows:

```cpp
template <unsigned int N>
struct factorial
{
    static constexpr unsigned int value =
        N * factorial<N - 1>::value;
};

template <>
struct factorial<0>
{
    static constexpr unsigned int value = 1;
};

int main()
{
    std::cout << factorial<4>::value << '\n';
}
```

The first definition is the primary template. It has a non-type template parameter representing the value whose factorial needs to be computed. This class contains a static constexpr data member called value, initialized with the result of multiplying the argument N and the value of the factorial class template instantiated with the decremented argument. The recursion needs an end case and that is provided by the explicit specialization for the value 0 (of the non-type template argument), in which case the member value is initialized with 1.

When encountering the instantiation factorial<4>::value in the main function, the compiler generates all the recursive instantiations from factorial<4> to factorial<0>. These look as follows:

```cpp
template<>
struct factorial<4>
{
    inline static constexpr const unsigned int value =
        4U * factorial<3>::value;
```

```
};

template<>
struct factorial<3>
{
    inline static constexpr const unsigned int value =
        3U * factorial<2>::value;
};

template<>
struct factorial<2>
{
    inline static constexpr const unsigned int value =
        2U * factorial<1>::value;
};

template<>
struct factorial<1>
{
    inline static constexpr const unsigned int value =
        1U * factorial<0>::value;
};

template<>
struct factorial<0>
{
    inline static constexpr const unsigned int value = 1;
};
```

From these instantiations, the compiler is able to compute the value of the data member `factorial<N>::value`. It should be mentioned again that when optimizations are enabled, this code would not even be generated, but the resulting constant is used directly in the generated assembly code.

The implementation of the factorial class template is relatively trivial, and the class template is basically only a wrapper over the static data member `value`. We can actually avoid it altogether by using a variable template instead. This can be defined as follows:

```
template <unsigned int N>
inline constexpr unsigned int factorial = N * factorial<N - 1>;

template <>
inline constexpr unsigned int factorial<0> = 1;

int main()
{
    std::cout << factorial<4> << '\n';
}
```

There is a striking similarity between the implementation of the `factorial` class template and the `factorial` variable template. For the latter, we have basically taken out the data member value and called it `factorial`. On the other hand, this may also be more convenient to use because it does not require accessing the data member value as in `factorial<4>::value`.

There is a third approach for computing the factorial at compile time: using function templates. A possible implementation is shown next:

```
template <unsigned int n>
constexpr unsigned int factorial()
{
    return n * factorial<n - 1>();
}

template<> constexpr unsigned int factorial<1>() {
                                        return 1; }
template<> constexpr unsigned int factorial<0>() {
                                        return 1; }

int main()
{
    std::cout << factorial<4>() << '\n';
}
```

You can see there is a primary template that calls the `factorial` function template recursively, and we have two full specializations for the values 1 and 0, both returning 1.

Which of these three different approaches is the best is probably arguable. Nevertheless, the complexity of the recursive instantiations of the factorial templates remained the same. However, this depends on the nature of the template. The following snippet shows an example of when complexity increases:

```cpp
template <typename T>
struct wrapper {};

template <int N>
struct manyfold_wrapper
{
    using value_type =
        wrapper<
            typename manyfold_wrapper<N - 1>::value_type>;
};

template <>
struct manyfold_wrapper<0>
{
    using value_type = unsigned int;
};

int main()
{
    std::cout <<
      typeid(manyfold_wrapper<0>::value_type).name() << '\n';
    std::cout <<
      typeid(manyfold_wrapper<1>::value_type).name() << '\n';
    std::cout <<
      typeid(manyfold_wrapper<2>::value_type).name() << '\n';
    std::cout <<
      typeid(manyfold_wrapper<3>::value_type).name() << '\n';
}
```

There are two class templates in this example. The first is called `wrapper` and has an empty implementation (it doesn't actually matter what it contains) but it represents a wrapper class over some type (or more precisely a value of some type). The second template is called `manyfold_wrapper`. This represents a wrapper over a wrapper over a type many times over, hence the name `manyfold_wrapper`. There is no end case for an upper limit of this number of wrappings, but there is a start case for the lower limit. The full specialization for value 0 defines a member type called `value_type` for the `unsigned int` type. As a result, `manyfold_wrapper<1>` defines a member type called `value_type` for `wrapper<unsigned int>`, `manyfold_wrapper<2>` defines a member type called `value_type` for `wrapper<wrapper<unsigned int>>`, and so on. Therefore, executing the `main` function will print the following to the console:

```
unsigned int
struct wrapper<unsigned int>
struct wrapper<struct wrapper<unsigned int> >
struct wrapper<struct wrapper<struct wrapper<unsigned int> > >
```

The C++ standard does not specify a limit for the recursively nested template instantiations but does recommend a minimum limit of 1,024. However, this is only a recommendation and not a requirement. Therefore, different compilers have implemented different limits. The **VC++ 16.11** compiler has the limit set at 500, **GCC 12** at 900, and **Clang 13** at 1,024. A compiler error is generated when this limit is exceeded. Some examples are shown here:

For VC++:

```
fatal error C1202: recursive type or function dependency
context too complex
```

For GCC:

```
fatal error: template instantiation depth exceeds maximum of
900 (use '-ftemplate-depth=' to increase the maximum)
```

For Clang:

```
fatal error: recursive template instantiation exceeded maximum
depth of 1024
use -ftemplate-depth=N to increase recursive template
instantiation depth
```

For GCC and Clang, the compiler option `-ftemplate-depth=N` can be used to increase this maximum value for nested template instantiations. Such an option is not available for the Visual C++ compiler.

Recursive templates help us solve some problems in a recursive manner at compile time. Whether you use recursive function templates, variable templates, or class templates depends on the problem you are trying to solve or perhaps your preference. However, you should keep in mind there are limits to the depth template recursion works. Nevertheless, use template recursion judiciously.

The next advanced topic to address in this chapter is template argument deduction, both for functions and classes. We start next with the former.

Function template argument deduction

Earlier in this book, we have briefly talked about the fact that the compiler can sometimes deduce the template arguments from the context of the function call, allowing you to avoid explicitly specifying them. The rules for template argument deduction are more complex and we will explore this topic in this section.

Let's start the discussion by looking at a simple example:

```
template <typename T>
void process(T arg)
{
    std::cout << "process " << arg << '\n';
}


int main()
{
    process(42);           // [1] T is int
    process<int>(42);      // [2] T is int, redundant
    process<short>(42);    // [3] T is short
}
```

In this snippet, `process` is a function template with a single type template parameter. The calls `process(42)` and `process<int>(42)` are identical because, in the first case, the compiler is able to deduce the type of the type template parameter `T` as `int` from the value of the argument passed to the function.

When the compiler tries to deduce the template arguments, it performs the matching of the types of the template parameters with the types of the arguments used to invoke the function. There are some rules that govern this matching. The compiler can match the following:

- Types (both cv-qualified and non-qualified) of the form `T`, `T const`, `T volatile`:

```
struct account_t
{
    int number;
};

template <typename T>
void process01(T) { std::cout << "T\n"; }

template <typename T>
void process02(T const) { std::cout << "T const\n"; }

template <typename T>
void process03(T volatile) { std::cout << "T volatile\n";
}

int main()
{
    account_t ac{ 42 };
    process01(ac);    // T
    process02(ac);    // T const
    process03(ac);    // T volatile
}
```

- Pointers (`T*`), l-value references (`T&`), and r-value references (`T&&`):

```
template <typename T>
void process04(T*) { std::cout << "T*\n"; }

template <typename T>
void process04(T&) { std::cout << "T&\n"; }

template <typename T>
```

```
void process05(T&&) { std::cout << "T&&\n"; }

int main()
{
    account_t ac{ 42 };
    process04(&ac);   // T*
    process04(ac);    // T&
    process05(ac);    // T&&
}
```

- Arrays such as `T[5]`, or `C[5][n]`, where `C` is a class type and `n` is a non-type template argument:

```
template <typename T>
void process06(T[5]) { std::cout << "T[5]\n"; }

template <size_t n>
void process07(account_t[5][n])
{ std::cout << "C[5][n]\n"; }

int main()
{
    account_t arr1[5] {};
    process06(arr1);   // T[5]

    account_t ac{ 42 };
    process06(&ac);    // T[5]

    account_t arr2[5][3];
    process07(arr2);   // C[5][n]
}
```

- Pointers to functions, with the form `T(*)()`, `C(*)(T)`, and `T(*)(U)`, where `C` is a class type and `T` and `U` are type template parameters:

```
template<typename T>
void process08(T(*)()) { std::cout << "T (*)()\n"; }
```

```cpp
template<typename T>
void process08(account_t(*)(T))
{ std::cout << "C (*)(T)\n"; }

template<typename T, typename U>
void process08(T(*)(U)) { std::cout << "T (*)(U)\n"; }

int main()
{
    account_t  (*pf1)()    = nullptr;
    account_t  (*pf2)(int) = nullptr;
    double     (*pf3)(int) = nullptr;

    process08(pf1);     // T (*)()
    process08(pf2);     // C (*)(T)
    process08(pf3);     // T (*)(U)
}
```

- Pointers to member functions with one of the following forms, `T (C::*)()`, `T (C::*)(U)`, `T (U::*)()`, `T (U::*)(V)`, `C (T::*)()`, `C (T::*)(U)`, and `D (C::*)(T)`, where `C` and `D` are class types and `T`, `U`, and `V` are type template parameters:

```cpp
struct account_t
{
    int number;

    int get_number() { return number; }
    int from_string(std::string text) {
        return std::atoi(text.c_str()); }
};

struct transaction_t
{
    double amount;
};
```

```
struct balance_report_t {};

struct balance_t
{
    account_t account;
    double    amount;

    account_t get_account()  { return account; }
    int get_account_number() { return account.number; }
    bool can_withdraw(double const value)
       {return amount >= value; };
    transaction_t withdraw(double const value) {
        amount -= value; return transaction_t{ -value }; }
    balance_report_t make_report(int const type)
    {return {}; }
};

template<typename T>
void process09(T(account_t::*)())
{ std::cout << "T (C::*)()\n"; }

template<typename T, typename U>
void process09(T(account_t::*)(U))
{ std::cout << "T (C::*)(U)\n"; }

template<typename T, typename U>
void process09(T(U::*)())
{ std::cout << "T (U::*)()\n"; }

template<typename T, typename U, typename V>
void process09(T(U::*)(V))
{ std::cout << "T (U::*)(V)\n"; }

template<typename T>
void process09(account_t(T::*)())
```

```cpp
{ std::cout << "C (T::*)()\n"; }

template<typename T, typename U>
void process09(transaction_t(T::*)(U))
{ std::cout << "C (T::*)(U)\n"; }

template<typename T>
void process09(balance_report_t(balance_t::*)(T))
{ std::cout << "D (C::*)(T)\n"; }

int main()
{
    int (account_t::* pfm1)() = &account_t::get_number;
    int (account_t::* pfm2)(std::string) =
        &account_t::from_string;
    int (balance_t::* pfm3)() =
        &balance_t::get_account_number;
    bool (balance_t::* pfm4)(double) =
        &balance_t::can_withdraw;
    account_t (balance_t::* pfm5)() =
        &balance_t::get_account;
    transaction_t(balance_t::* pfm6)(double) =
        &balance_t::withdraw;
    balance_report_t(balance_t::* pfm7)(int) =
        &balance_t::make_report;

    process09(pfm1);     // T (C::*)()
    process09(pfm2);     // T (C::*)(U)
    process09(pfm3);     // T (U::*)()
    process09(pfm4);     // T (U::*)(V)
    process09(pfm5);     // C (T::*)()
    process09(pfm6);     // C (T::*)(U)
    process09(pfm7);     // D (C::*)(T)
}
```

- Pointers to data members such as `T C::*`, `C T::*`, and `T U::*`, where `C` is a class type and `T` and `U` are type template parameters:

```
template<typename T>
void process10(T account_t::*)
{ std::cout << "T C::*\n"; }


template<typename T>
void process10(account_t T::*)
{ std::cout << "C T::*\n"; }


template<typename T, typename U>
void process10(T U::*) { std::cout << "T U::*\n"; }


int main()
{
    process10(&account_t::number);    // T C::*
    process10(&balance_t::account);   // C T::*
    process10(&balance_t::amount);    // T U::*
}
```

- A template with an argument list that contains at least one type template parameter; the general form is `C<T>`, where `C` is a class type and `T` is a type template parameter:

```
template <typename T>
struct wrapper
{
    T data;
};


template<typename T>
void process11(wrapper<T>) { std::cout << "C<T>\n"; }


int main()
{
    wrapper<double> wd{ 42.0 };
    process11(wd); // C<T>
}
```

- A template with an argument list that contains at least one non-type template argument; the general form is `C<i>`, where `C` is a class type and `i` a non-type template argument:

```
template <size_t i>
struct int_array
{
    int data[i];
};

template<size_t i>
void process12(int_array<i>) { std::cout << "C<i>\n"; }

int main()
{
    int_array<5> ia{};
    process12(ia); // C<i>
}
```

- A template template argument with an argument list that contains at least one type template parameter; the general form is `TT<T>`, where `TT` is a template template parameter and `T` is a type template:

```
template<template<typename> class TT, typename T>
void process13(TT<T>) { std::cout << "TT<T>\n"; }

int main()
{
    wrapper<double> wd{ 42.0 };
    process13(wd);      // TT<U>
}
```

- A template template argument with an argument list that contains at least one non-type template argument; the general form is `TT<i>`, where `TT` is a template template parameter and `i` is a non-type template argument:

```
template<template<size_t> typename TT, size_t i>
void process14(TT<i>) { std::cout << "TT<i>\n"; }
int main()
```

```
{
    int_array<5> ia{};
    process14(ia);      // TT<i>
}
```

- A template template argument with an argument list that has no template arguments dependent on a template parameter; this has the form TT<C>, where TT is the template template parameter and C is a class type:

```
template<template<typename> typename TT>
void process15(TT<account_t>) { std::cout << "TT<C>\n"; }

int main()
{
    wrapper<account_t> wa{ {42} };
    process15(wa);      // TT<C>
}
```

Although the compiler is able to deduce many types of template parameters, as previously seen, there are also limitations to what it can do. These are exemplified in the following list:

- The compiler cannot deduce the type of a type template argument, from the type of a non-type template argument. In the following example, process is a function template with two template parameters: a type template called T, and a non-type template i of the type T. Calling the function with an array of five doubles does not allow the compiler to determine the type of T, even though this is the type of the value specifying the size of the array:

```
template <typename T, T i>
void process(double arr[i])
{
    using index_type = T;
    std::cout << "processing " << i
              << " doubles" << '\n';

std::cout << "index type is "
          << typeid(T).name() << '\n';
}

int main()
```

```
{
    double arr[5]{};
    process(arr);           // error
    process<int, 5>(arr);   // OK
}
```

- The compiler is not able to determine the type of a template argument from the type of a default value. This is exemplified ahead in the code with the function template process, which has a single type template parameter, but two function parameters, both of type T and both with default values.

The process() call (without any arguments) fails because the compiler cannot deduce the type of the type template parameter T from the default values of the function parameters. The process<int>() call is OK because the template argument is explicitly provided. The process(6) call is also OK, because the type of the first function parameter can be deduced from the supplied argument, and, therefore, the type template argument can also be deduced:

```
template <typename T>
void process(T a = 0, T b = 42)
{
    std::cout << a << "," << b << '\n';
}

int main()
{
    process();           // [1] error
    process<int>();      // [2] OK
    process(10);         // [3] OK
}
```

- Although the compiler can deduce function template arguments from pointer to functions or pointer to member functions, as we have seen earlier, there are a couple of restrictions to this capability: it cannot deduce arguments from pointers to function templates, nor from a pointer to a function that has an overloaded set with more than one overloaded function matching the required type.

In the code ahead, the function template `invoke` takes a pointer to a function that has two arguments, the first of the type template parameter `T`, and the second an `int`, and returns `void`. This function template cannot be passed a pointer to `alpha` (see [1]) because this is a function template, nor to `beta` (see [2]), because this has more than one overload that can match the type `T`. However, it is possible to call it with a pointer to `gamma` (see [3]), and it will correctly deduce the type of the second overload:

```cpp
template <typename T>
void invoke(void(*pfun)(T, int))
{
    pfun(T{}, 42);
}

template <typename T>
void alpha(T, int)
{ std::cout << "alpha(T,int)" << '\n'; }

void beta(int, int)
{ std::cout << "beta(int,int)" << '\n'; }
void beta(short, int)
{ std::cout << "beta(short,int)" << '\n'; }

void gamma(short, int, long long)
{ std::cout << "gamma(short,int,long long)" << '\n'; }
void gamma(double, int)
{ std::cout << "gamma(double,int)" << '\n'; }

int main()
{
    invoke(&alpha);   // [1] error
    invoke(&beta);    // [2] error
    invoke(&gamma);   // [3] OK
}
```

- Another limitation of the compiler is the argument deduction of the primary dimension of an array. The reason is this is not part of function parameter types. The exceptions to this limitation are the cases when the dimension refers to a reference or pointer type. The following code snippet demonstrates these restrictions:

  - The call to process1() at [1] generates an error because the compiler is not able to deduce the value of the non-type template argument Size, since this refers to the primary dimension of an array.

  - The call to process2() at the point marked with [2] is correct because the non-type template parameter Size refers to the second dimension of an array.

  - On the other hand, the calls to process3() (at [3]) and process4() (at [4]) are both successful, since the function argument is either a reference or a pointer to a single-dimensional array:

```
template <size_t Size>
void process1(int a[Size])
{ std::cout << "process(int[Size])" << '\n'; };

template <size_t Size>
void process2(int a[5][Size])
{ std::cout << "process(int[5][Size])" << '\n'; };

template <size_t Size>
void process3(int (&a)[Size])
{ std::cout << "process(int[Size]&)" << '\n'; };

template <size_t Size>
void process4(int (*a)[Size])
{ std::cout << "process(int[Size]*)" << '\n'; };

int main()
{
    int arr1[10];
    int arr2[5][10];

    process1(arr1);    // [1] error
    process2(arr2);    // [2] OK
    process3(arr1);    // [3] OK
```

```
        process4(&arr1);   // [4] OK
    }
```

- If a non-type template argument is used in an expression in the function template parameter list, then the compiler cannot deduce its value.

  In the following snippet, ncube is a class template with a non-type template parameter N representing a number of dimensions. The function template process also has a non-type template parameter N, but this is used in an expression in the template parameter list of the type of its single parameter. As a result, the compiler cannot deduce the value of N from the type of the function argument (as seen at [1]) and this must be specified explicitly (as seen at [2]):

```
template <size_t N>
struct ncube
{
    static constexpr size_t dimensions = N;
};


template <size_t N>
void process(ncube<N - 1> cube)
{
    std::cout << cube.dimensions << '\n';
}


int main()
{
    ncube<5> cube;
    process(cube);      // [1] error
    process<6>(cube);  // [2] OK
}
```

All the rules for template argument deduction discussed in this section also apply to variadic function templates. However, everything that was discussed was in the context of function templates. Template argument deduction works for class templates too and we will explore this topic in the next section.

Class template argument deduction

Before **C++17**, template argument deduction only worked for functions but not classes. This meant that when a class template had to be instantiated, all the template arguments had to be supplied. The following snippet shows several examples:

```
template <typename T>
struct wrapper
{
    T data;
};


std::pair<int, double> p{ 42, 42.0 };
std::vector<int>       v{ 1,2,3,4,5 };
wrapper<int>           w{ 42 };
```

By leveraging template argument deduction for function templates, some standard types feature helper functions that create an instance of the type without the need to explicitly specify template arguments. Such examples are std::make_pair for std::pair and std::make_unique for std::unique_ptr. These helper function templates, used in corroboration with the auto keyword, avoid the need for specifying template arguments for class templates. Here is an example:

```
auto p = std::make_pair(42, 42.0);
```

Although not all standard class templates have such a helper function for creating instances, it's not hard to write your own. In the following snippet, we can see a make_vector function template used to create a std::vector<T> instance, and a make_wrapper function template to create a wrapper<T> instance:

```
template <typename T, typename... Ts,
          typename Allocator = std::allocator<T>>
auto make_vector(T&& first, Ts&&... args)
{
    return std::vector<std::decay_t<T>, Allocator> {
        std::forward<T>(first),
        std::forward<Ts>(args)...
    };
}
```

```
template <typename T>
constexpr wrapper<T> make_wrapper(T&& data)
{
    return wrapper{ data };
}


auto v = make_vector(1, 2, 3, 4, 5);
auto w = make_wrapper(42);
```

The C++17 standard has simplified the use of class templates by providing template argument deduction for them. Therefore, as of C++17, the first snippet shown in this section can be simplified as follows:

```
std::pair    p{ 42, 42.0 };    // std::pair<int, double>
std::vector v{ 1,2,3,4,5 };    // std::vector<int>
wrapper     w{ 42 };           // wrapper<int>
```

This is possible because the compiler is able to deduce the template arguments from the type of the initializers. In this example, the compiler deduced it from the initializing expression of the variables. But the compiler is also able to deduce template arguments from new expressions and function-style cast expressions. These are exemplified next:

```
template <typename T>
struct point_t
{
    point_t(T vx, T vy) : x(vx), y(vy) {}
private:
    T x;
    T y;
};


auto p = new point_t(1, 2);    // [1] point<int>
                               // new expression

std::mutex mt;
auto l = std::lock_guard(mt); // [2]
// std::lock_guard<std::mutex>
// function-style cast expression
```

The way template argument deduction works for class templates is different than for function templates but it relies on the latter. When encountering the name of a class template in a variable declaration or function-style cast, the compiler proceeds to build a set of so-called **deduction guides**.

There are fictional function templates representing constructor signatures of a *fictional class type*. Users can also provide deduction guides and these are added to the list of compiler-generated guides. If overload resolution fails on the constructed set of fictional function templates (the return type is not part of the matching process since these functions represent constructors), then the program is ill-formed and an error is generated. Otherwise, the return type of the selected function template specialization becomes the deduced class template specialization.

To understand this better, let's see how the deduction guides actually look. In the following snippet, you can see some of the guides generated by the compiler for the `std::pair` class. The actual list is longer and, for brevity, only some are presented here:

```
template <typename T1, typename T2>
std::pair<T1, T2> F();


template <typename T1, typename T2>
std::pair<T1, T2> F(T1 const& x, T2 const& y);


template <typename T1, typename T2, typename U1,
          typename U2>
std::pair<T1, T2> F(U1&& x, U2&& y);
```

This set of implicitly deduced guides is generated from the constructors of the class template. This includes the default constructor, the copy constructor, the move constructor, and all the conversion constructors, with the arguments copied in the exact order. If the constructor is explicit, then so is the deduction guide. However, if the class template does not have any user-defined constructor, a deduction guide is created for a hypothetical default constructor. A deduction guide for a hypothetical copy constructor is always created.

User-defined deduction guides can be provided in the source code. The syntax is similar to that of functions with a trailing return type but without the `auto` keyword. Deduction guides can be either functions or function templates. What is important to keep in mind is that these must be provided in the same namespace as the class template they apply to. Therefore, if we were to add a user-defined deduction guide for the `std::pair` class, it must be done in the `std` namespace. An example is shown here:

```
namespace std
{
    template <typename T1, typename T2>
    pair(T1&& v1, T2&& v2) -> pair<T1, T2>;
}
```

The deduction guides shown so far were all function templates. But as mentioned earlier, they don't have to be function templates. They can be regular functions too. To demonstrate this, let's consider the following example:

```
std::pair  p1{1, "one"};    // std::pair<int, const char*>
std::pair  p2{"two", 2};    // std::pair<const char*, int>
std::pair  p3{"3", "three"};
                    // std::pair<const char*, const char*>
```

With the compiler-degenerated deduction guides for the `std::pair` class, the deduced types are `std::pair<int, const char*>` for p1, `std::pair<const char*, int>` for p2, and `std::pair<const char*, const char*>` for p3. In other words, the type deduced by the compiler where literal strings are used is `const char*` (as one should expect). We could tell the compiler to deduce `std::string` instead of `const char*` by providing several user-defined deduction guides. These are shown in the following listing:

```
namespace std
{
    template <typename T>
    pair(T&&, char const*) -> pair<T, std::string>;

    template <typename T>
    pair(char const*, T&&) -> pair<std::string, T>;
```

```
    pair(char const*, char const*) ->
        pair<std::string, std::string>;
}
```

Notice that the first two are function templates, but the third one is a regular function. Having these guides available, the deduced types for p1, p2, and p3 from the previous example are std::pair<int, std::string>, std::pair<std::string, int> and std::pair<std::string, std::string> respectively.

Let's look at one more example for user-defined guides, this time for a user-defined class. Let's consider the following class template that models a range:

```
template <typename T>
struct range_t
{
    template <typename Iter>
    range_t(Iter first, Iter last)
    {
        std::copy(first, last, std::back_inserter(data));
    }
private:
    std::vector<T> data;
};
```

There is not much to this implementation but, in fact, it is enough for our purpose. Let's consider you want to construct a range object from an array of integers:

```
int arr[] = { 1,2,3,4,5 };
range_t r(std::begin(arr), std::end(arr));
```

Running this code will generate an error. Different compilers would generate different error messages. Perhaps Clang provides the error messages that best describe the problem:

```
error: no viable constructor or deduction guide for deduction
of template arguments of 'range_t'
    range_t r(std::begin(arr), std::end(arr));
            ^

note: candidate template ignored: couldn't infer template
argument 'T'
        range_t(Iter first, Iter last)
```

```
                 ^

note: candidate function template not viable: requires 1
argument, but 2 were provided
    struct range_t
```

Nevertheless, regardless of what the actual error message is, the meaning is the same: template argument deduction for range_t failed. In order to make deduction work, a user-defined deduction guide needs to be provided and it needs to look as follows:

```
template <typename Iter>
range_t(Iter first, Iter last) ->
    range_t<
        typename std::iterator_traits<Iter>::value_type>;
```

What this deduction guide is instructing is that when a call to the constructor with two iterator arguments is encountered, the value of the template parameter T should be deduced to be the value type of the iterator traits. Iterator traits is a topic that will be addressed in *Chapter 5, Type Traits and Conditional Compilation*. However, with this available, the previous snippet runs without problems and the compiler deduces the type of the r variable to be range_t<int>, as intended.

At the beginning of this section, the following example was provided, where the type of w was said to be deduced as wrapper<int>:

```
wrapper w{ 42 }; // wrapper<int>
```

In C++17, this is not actually true without a user-defined deduction guide. The reason is that wrapper<T> is an aggregate type and class template argument deduction does not work from aggregate initialization in C++17. Therefore, to make the preceding line of code work, a deduction guide as follows needs to be provided:

```
template <typename T>
wrapper(T) -> wrapper<T>;
```

Fortunately, the need for such a user-defined deduction guide was removed in C++20. This version of the standard provides support for aggregate types (as long as any dependent base class has no virtual functions or virtual base classes and the variable is initialized from a non-empty list of initializers).

Class template argument deduction only works *if no template arguments are provided*. As a consequence, the following declarations of p1 and p2 are both valid and class template argument deduction occurs; for p2, the deduced type is std::pair<int, std::string> (assuming the previously user-defined guides are available). However, the declarations of p3 and p4 produce an error because class template argument deduction does not occur, since a template argument list is present (`<>` and `<int>`) but does not contain all required arguments:

```
std::pair<int, std::string> p1{ 1, "one" };   // OK
std::pair p2{ 2, "two" };                      // OK
std::pair<> p3{ 3, "three" };                  // error
std::pair<int> p4{ 4, "four" };                // error
```

Class template argument deduction may not always produce the expected results. Let's consider the following example:

```
std::vector v1{ 42 };
std::vector v2{ v1, v1 };
std::vector v3{ v1 };
```

The deduced type for v1 is `std::vector<int>` and the deduced type for v2 is `std::vector<std::vector<int>>`. However, what should the compiler deduce for the type of v3? There are two options: `std::vector<std::vector<int>>` and `std::vector<int>`. If your expectation is the former, you will be disappointed to learn that the compiler actually deduces the latter. This is because deduction depends on both *the number of arguments and their type*.

When the number of arguments is greater than one, it will use the constructor that takes an initializer list. For the v2 variable, that is `std::initializer_list<std::vector<int>>`. When the number of arguments is one, then the type of the arguments is considered. If the type of the argument is a (specialization of) `std::vector` – considering this explicit case – then the copy-constructor is used and the deduced type is the declared type of the argument. This is the case of variable v3, where the deduced type is `std::vector<int>`. Otherwise, the constructor that takes an initializer list (with a single element) is used, as in the case of variable v1, for which the deduced type is `std::vector<int>`. These could be better visualized with the help of the `cppinsights.io` tool, which shows the following generated code (for the previous snippet). Notice that the allocator argument has been removed for brevity:

```
std::vector<int> v1 =
    std::vector<int>{std::initializer_list<int>{42}};

std::vector<vector<int>> v2 =
    std::vector<vector<int>>{
        std::initializer_list<std::vector<int>>{
            std::vector<int>(v1),
```

```
            std::vector<int>(v1)
        }
    };
```

```
std::vector<int> v3 = std::vector<int>{v1};
```

Class template argument deduction is a useful feature of C++17 with improvements for aggregate types in C++20. It helps avoid writing unnecessary explicit template arguments when the compiler is able to deduce them, even though, in some cases, the compiler may require user-defined deduction guides for the deduction to work. It also avoids the need for creating factory functions, such as `std::make_pair` or `std::make_tuple`, that were a workaround for benefiting from template argument deduction before it was available for class templates.

There is more to template argument deduction than what we have discussed so far. There is a special case of function template argument deduction known as forwarding references. This will be addressed next.

Forwarding references

One of the most important features that were added to the language in C++11 is **move semantics**, which helps improve performance by avoiding making unnecessary copies. Move semantics are supported by another C++11 feature called **rvalue references**. Before discussing these, it is worth mentioning that, in C++, we have two kinds of values:

- **lvalues** are values that refer to a memory location and, therefore, we can take their address with the & operator. lvalues can appear both on the left and right sides of an assignment expression.

- **rvalues** are values that are not lvalues. They are defined by exclusion. rvalues do not refer to a memory location and you can't take their address with the & operator. rvalues are literals and temporary objects and can only appear on the right side of an assignment expression.

> **Note**
>
> In C++11, there are a few other value categories, glvalue, prvalue, and xvalue. Discussing them here would not benefit the current topic. However, you can read more about them at `https://en.cppreference.com/w/cpp/language/value_category`.

References are aliases to already existing objects or functions. Just as we have two kinds of values, in C++11 we have two kinds of references:

- **lvalue references**, denoted with a &, such as in &x, are references to lvalues.

- **rvalue references**, denoted with &&, such as in &&x, are references to rvalues.

Let's look at some examples to understand these concepts better:

```cpp
struct foo
{
    int data;
};

void f(foo& v)
{ std::cout << "f(foo&)\n"; }

void g(foo& v)
{ std::cout << "g(foo&)\n"; }

void g(foo&& v)
{ std::cout << "g(foo&&)\n"; }

void h(foo&& v)
{ std::cout << "h(foo&&)\n"; }

foo x = { 42 };    //  x is lvalue
foo& rx = x;       // rx is lvalue
```

We have three functions here: f, which takes an lvalue reference (that is, int &); g, which has two overloads, one for an lvalue reference, and one for an rvalue reference (that is, int &&); and h, which takes an rvalue reference. We also have two variables, x and rx. Here, x is an lvalue, whose type is foo. We can take its address with &x. An lvalue is also rx, which is an lvalue reference, whose type is foo&. Now, let's see how we can call each of the f, g, and h functions:

```cpp
f(x);        // f(foo&)
f(rx);       // f(foo&)
f(foo{42}); // error: a non-const reference
             // may only be bound to an lvalue
```

Because x and rx are both lvalues, passing them to f is OK since this function takes an lvalue reference. However, foo{42} is a temporary object, as it does not exist outside the context of the call to f. That means, it is an rvalue, and passing it to f will result in a compiler error, because the parameter of the function is of the type foo& and non-constant references may only be bound to lvalues. This would work if the signature of the function f was changed to f(int const &v). Next, let's discuss the g function:

```
g(x);               // g(foo&)
g(rx);              // g(foo&)
g(foo{ 42 });       // g(foo&&)
```

In the preceding snippet, calling g with either x or rx will resolve to the first overload, which takes an lvalue reference. However, calling it with foo{42}, which is a temporary object, therefore an rvalue, will resolve to the second overload, which takes an rvalue reference. Let's see what happens when we want to make the same calls to the h function:

```
h(x);           // error, cannot bind an lvalue to an rvalue ref
h(rx);              // error
h(foo{ 42 });       // h(foo&&)
h(std::move(x)); // h(foo&&)
```

This function takes an rvalue reference. The attempts to pass either x or rx to it result in compiler errors because lvalues cannot be bound to rvalue references. The expression foo{42}, being an rvalue, can be passed as an argument. We can also pass the lvalue x to the function h if we change its semantic from an lvalue to an rvalue. That is done with the help of std::move. This function does not really move anything; it only makes a sort of a cast from an lvalue to an rvalue.

However, it is important to understand that passing rvalues to a function has two purposes: either the object is temporary and does not exist outside the call and the function can do anything with it, or the function is supposed to take ownership of the object that is received. This is the purpose of the move constructor and the move assignment operator and it's rare that you will see other functions taking rvalue references. In our last example, within the function h, the parameter v is an lvalue but it is bound to an rvalue. The variable x existed outside the call to h but passing it to std::move transformed it into an rvalue. It still exists as an lvalue after the call to h returns but you should assume the function h did something with it and its state can be anything.

One purpose of rvalue references is, therefore, to enable move semantics. But it has yet another one and that is to enable **perfect forwarding**. To understand this, let's consider the following modified scenario of the previous functions g and h:

```
void g(foo& v)   { std::cout << "g(foo&)\n"; }
void g(foo&& v)  { std::cout << "g(foo&&)\n"; }


void h(foo& v)   { g(v); }
void h(foo&& v)  { g(v); }
```

In this snippet, the implementation of g is identical to the one seen earlier. However, h also has two overloads, one that takes an lvalue reference and calls g and another one that takes an rvalue reference and also calls g. In other words, the function h is just forwarding the argument to g. Now, let's consider the following calls:

```
foo x{ 42 };
h(x);            // g(foo&)
h(foo{ 42 });    // g(foo&)
```

From this, you would expect that the call h(x) will result in a call to the g overload taking an lvalue reference and the call to h(foo{42}) will result in a call to the g overload taking an rvalue reference. However, in fact, both of them will call the first overload of g, therefore printing g(foo&) to the console. The explanation is actually simple once you understand how references work: in the context h(foo&& v), the parameter v is actually an lvalue (it has a name and you can take its address) so calling g with it invokes the overload that takes an lvalue reference. To make it work as intended, we need to change the implementation of the h functions as follows:

```
void h(foo& v)   { g(std::forward<foo&>(v)); }
void h(foo&& v)  { g(std::forward<foo&&>(v)); }
```

The std::forward is a function that enables the correct forwarding of values. What the function does is as follows:

- If the argument is an lvalue reference, then the function behaves just as a call to std::move (changing the semantics from an lvalue to an rvalue).

- If the argument is an rvalue reference, then it does nothing.

Everything that we discussed so far is unrelated to templates, which are the subject of this book. However, function templates can also be used to take lvalue and rvalue references and it's important to understand first how these work in non-templates scenarios. This is because, in templates, rvalue references work slightly differently, and sometimes they are rvalue references, but other times they are actually lvalue references.

References that exhibit this behavior are called **forwarding references**. However, they are often referred to as **universal references**. This was a term coined by Scott Meyers shortly after C++11 when there was no term in the standard for this type of reference. In order to address this shortcoming, and because it didn't feel the term universal references properly described their semantics, the C++ standard committee called these forwarding references in C++14. Yet, both terms are equally present in literature. For the sake of being true to the standard terminology, we'll call them forwarding references in this book.

To begin the discussion on forwarding references, let's consider the following overloaded function templates and class templates:

```
template <typename T>
void f(T&& arg)                    // forwarding reference
{ std::cout << "f(T&&)\n"; }


template <typename T>
void f(T const&& arg)              // rvalue reference
{ std::cout << "f(T const&&)\n"; }


template <typename T>
void f(std::vector<T>&& arg)    // rvalue reference
{ std::cout << "f(vector<T>&&)\n"; }


template <typename T>
struct S
{
    void f(T&& arg)                // rvalue reference
    { std::cout << "S.f(T&&)\n"; }
};
```

We can make calls to these functions as follows:

```
int x = 42;
f(x);                        // [1] f(T&&)
```

```
f(42);                      // [2] f(T&&)

int const cx = 100;
f(cx);                      // [3] f(T&&)
f(std::move(cx));           // [4] f(T const&&)

std::vector<int> v{ 42 };
f(v);                       // [5] f(T&&)
f(std::vector<int>{42});;// [6] f(vector<T>&&)

S<int> s;
s.f(x);                     // [7] error
s.f(42);                    // [8] S.f(T&&)
```

From this snippet, we can notice that:

- Calling f with an lvalue or rvalue at [1] and [2] resolves to the first overload, f(T&&).

- Calling f with a constant lvalue at [3] also resolves to the first overload, but calling f with a constant rvalue at [4] resolves to the second overload, f(T const&&), because it's a better match.

- Calling f with an lvalue std::vector object at [5] resolves to the first overload, but calling f with an rvalue std::vector object at [6] resolves to the third overload, f(vector<T>&&), because it's a better match.

- Calling S::f with an lvalue at [7] is an error because lvalues cannot be bound to rvalue references, but calling it with an rvalue at [8] is correct.

All the f function overloads in this example take an rvalue reference. However, the && in the first overload does not necessarily mean an rvalue reference. It means *an rvalue reference if an rvalue was passed or an lvalue reference if an lvalue was passed.* Such a reference is called a **forwarding reference**. However, *forwarding references are only present in the context of an rvalue reference to a template parameter.* It has to have the form T&& and nothing else. T const&& or std::vector<T>&& are not forwarding references, but normal rvalue references. Similarly, the T&& in the f function member of the class template S is also an rvalue reference because f is not a template but a non-template member function of a class template, so this rule for forwarding references does not apply.

Forwarding references are a special case of function template argument deduction, a topic that we previously discussed in this chapter. Their purpose is to enable perfect forwarding with templates and they are made possible by a new C++11 feature called **reference collapsing**. Let's look at this first, before showing how they solve the perfect forwarding problem.

Prior to C++11, it was not possible to take a reference to a reference. However, that is now possible in C++11 for typedefs and templates. Here is an example:

```
using lrefint = int&;
using rrefint = int&&;
int x = 42;
lrefint&  r1 = x; // type of r1 is int&
lrefint&& r2 = x; // type of r2 is int&
rrefint&  r3 = x; // type of r3 is int&
rrefint&& r4 = 1; // type of r4 is int&&
```

The rule is pretty simple: an rvalue reference to an rvalue reference collapses to an rvalue reference; all other combinations collapse to an lvalue reference. This can be put in a tabular form as follows:

Type	Type of reference	Type of variable
T&	T&	T&
T&	T&&	T&
T&&	T&	T&
T&&	T&&	T&&

Table 4.2

Any other combinations, shown in the following table, do not involve reference collapsing rules. These only apply when both types are references:

Type	Type of reference	Type of variable
T	T	T
T	T&	T&
T	T&&	T&&
T&	T	T&
T&&	T	T&&

Table 4.3

Forwarding references work not only for templates but also with auto deduction rules. When `auto&&` is found, it means a forwarding reference. The same does not apply for anything else, such as cv-qualified forms like `auto const&&`. Here are some examples:

```
int x = 42;
auto&& rx = x;              // [1] int&
auto&& rc = 42;            // [2] int&&
auto const&& rcx = x;     // [3] error

std::vector<int> v{ 42 };
auto&& rv = v[0];         // [4] int&
```

In the first two examples, `rx` and `rc` are both forwarding references and are bound to an lvalue and an rvalue respectively. However, `rcx` is an rvalue reference because `auto const&&` does not denote a forwarding reference. Therefore, trying to bind it to an lvalue is an error. Similarly, `rv` is a forwarding reference and is bound to an lvalue.

As previously mentioned, the purpose of forwarding references is to enable perfect forwarding. We have seen the concept of perfect forwarding earlier but in a non-template context. It works, however, in a similar manner with templates. To demonstrate this, let's redefine the function h as a template function. It would look as follows:

```
void g(foo& v)  { std::cout << "g(foo&)\n"; }
void g(foo&& v) { std::cout << "g(foo&&)\n"; }

template <typename T> void h(T& v)  { g(v); }
template <typename T> void h(T&& v) { g(v); }

foo x{ 42 };
h(x);          // g(foo&)
h(foo{ 42 });  // g(foo&)
```

The implementation of the g overloads is the same, but the h overloads are now function templates. However, calling h with an lvalue and an rvalue actually resolves to the same call to g, the first overload taking an lvalue. This is because in the context of the function h, v is an lvalue so passing it to g will call the overload taking an lvalue.

The solution to this problem is the same as what we already saw before discussing templates. However, there is a difference: we no longer need two overloads, but a single one taking a forwarding reference:

```
template <typename T>
void h(T&& v)
{
    g(std::forward<T>(v));
}
```

This implementation is using `std::forward` to pass lvalues as lvalues and rvalues as rvalues. It works similarly for variadic function templates. The following is a conceptual implementation of the `std::make_unique` function that creates a `std::unique_ptr` object:

```
template<typename T, typename... Args>
std::unique_ptr<T> make_unique(Args&&... args)
{
    return std::unique_ptr<T>(
            new T(std::forward<Args>(args)...));
}
```

To summarize this section, remember that forwarding references (also known as **universal references**) are basically a special deduction rule for function template arguments. They work based on the rules of reference collapsing and their purpose is to enable perfect forwarding. That is passing forward to another function a reference by preserving its valueness semantics: rvalues should be passed as rvalues and lvalues as lvalues.

The next topic that we will address in this chapter is the `decltype` specifier.

The decltype specifier

This specifier, introduced in C++11, returns the type of an expression. It is usually used in templates together with the `auto` specifier. Together, they can be used to declare the return type of a function template that depends on its template arguments, or the return type of a function that wraps another function and returns the result from executing the wrapped function.

The `decltype` specifier is not restricted for use in template code. It can be used with different expressions, and it yields different results based on the expression. The rules are as follows:

1. If the expression is an identifier or a class member access, then the result is the type of the entity that is named by the expression. If the entity does not exist, or it is a function that has an overload set (more than one function with the same name exists), then the compiler will generate an error.

2. If the expression is a function call or an overloaded operator function, then the result is the return type of the function. If the overloaded operator is wrapped in parentheses, these are ignored.

3. If the expression is an lvalue, then the result type is an lvalue reference to the type of expression.

4. If the expression is something else, then the result type is the type of the expression.

To understand these rules better, we'll look at a set of examples. For these, we will consider the following functions and variables that we will use in `decltype` expressions:

```
int f() { return 42; }
int g() { return 0; }
int g(int a) { return a; }

struct wrapper
{
    int val;
    int get() const { return val; }
};

int a = 42;
int& ra = a;
const double d = 42.99;
long arr[10];
long l = 0;
char* p = nullptr;
char c = 'x';
wrapper w1{ 1 };
wrapper* w2 = new wrapper{ 2 };
```

The following listing shows multiple uses of the `decltype` specifier. The rule that applies in each case, as well as the deduced type, is specified on each line in a comment:

```
decltype(a) e1;               // R1, int
decltype(ra) e2 = a;          // R1, int&
decltype(f) e3;               // R1, int()
decltype(f()) e4;             // R2, int
decltype(g) e5;               // R1, error
decltype(g(1)) e6;            // R2, int
decltype(&f) e7 = nullptr;    // R4, int(*)()
decltype(d) e8 = 1;           // R1, const double
decltype(arr) e9;             // R1, long[10]
decltype(arr[1]) e10 = 1;     // R3, long&
decltype(w1.val) e11;         // R1, int
decltype(w1.get()) e12;       // R1, int
decltype(w2->val) e13;        // R1, int
decltype(w2->get()) e14;      // R1, int
decltype(42) e15 = 1;         // R4, int
decltype(1 + 2) e16;          // R4, int
decltype(a + 1) e17;          // R4, int
decltype(a = 0) e18 = a;      // R3, int&
decltype(p) e19 = nullptr;    // R1, char*
decltype(*p) e20 = c;         // R3, char&
decltype(p[0]) e21 = c;       // R3, char&
```

We will not elaborate on all these declarations. Most of these are relatively easy to follow based on the specified rules. A few notes, however, are worth considering for clarifying some of the deduced types:

- `decltype(f)` only names a function with an overloaded set, so rule 1 applies. `decltype(g)` also names a function but it has an overloaded set. Therefore, rule 1 applies and the compiler generates an error.

- `decltype(f())` and `decltype(g(1))` are both using function calls for the expression, so the second rule applies, and even if g has an overload set, the declaration is correct.

- `decltype(&f)` uses the address of the function f, so the fourth rule applies, yielding `int(*)()`.

- `decltype(1+2)` and `decltype(a+1)` use the overloaded operator +
  that returns an rvalue, so the fourth rule applies. The result is `int`. However,
  `decltype(a = 1)` uses the assignment operator that returns an lvalue, so the
  third rule applies, yielding the lvalue reference `int&`.

The `decltype` specifier defines an **unevaluated context**. This means the expression used
with the specifier is not evaluated since this specifier is only querying the properties of its
operand. You can see this in the following snippet, where the assignment `a=1` is used with
the `decltype` specifier to declare the variable `e`, but after the declaration, the value of
`a` is the one with which it was initialized:

```
int a = 42;
decltype(a = 1) e = a;
std::cout << a << '\n';   // prints 42
```

There is an exception to this rule concerning template instantiation. When the expression
used with the `decltype` specifier contains a template, the template is instantiated before
the expression is evaluated at compile time:

```
template <typename T>
struct wrapper
{
    T data;
};

decltype(wrapper<double>::data) e1;   // double

int a = 42;
decltype(wrapper<char>::data, a) e2; // int&
```

The type of `e1` is `double`, and `wrapper<double>` is instantiated for this to be
deduced. On the other hand, the type of `e2` is `int&` (as the variable `a` is an lvalue).
However, `wrapper<char>` is instantiated here even if the type is only deduced from
the variable `a` (due to the use of the comma operator).

The preceding rules mentioned are not the only ones used for determining the type. There are several more for data member access. These are as follows:

- The `const` or `volatile` specifiers of the object used in the `decltype` expression do not contribute to the deduced type.

- Whether the object or pointer expression is an lvalue or an rvalue does not affect the deduced type.

- If the data member access expression is parenthesized, such as `decltype((expression))`, then the previous two rules do not apply. The `const` or `volatile` qualifier of the object does affect the deduced type, including the valueness of the object.

The first two rules from this list are demonstrated with the following snippet:

```
struct foo
{
    int        a = 0;
    volatile int b = 0;
    const int    c = 42;
};

foo f;
foo const cf;
volatile foo* pf = &f;

decltype(f.a) e1 = 0;           // int
decltype(f.b) e2 = 0;           // int volatile
decltype(f.c) e3 = 0;           // int const

decltype(cf.a) e4 = 0;          // int
decltype(cf.b) e5 = 0;          // int volatile
decltype(cf.c) e6 = 0;          // int const

decltype(pf->a) e7 = 0;         // int
decltype(pf->b) e8 = 0;         // int volatile
decltype(pf->c) e9 = 0;         // int const

decltype(foo{}.a) e10 = 0;  // int
```

```
decltype(foo{}.b) e11 = 0;   // int volatile
decltype(foo{}.c) e12 = 0;   // int const
```

The deduced type for each case is mentioned on the right side in a comment. When the expression is parenthesized, these two rules are reversed. Let's take a look at the following snippet:

```
foo f;
foo const cf;
volatile foo* pf = &f;

int x = 1;
int volatile y = 2;
int const z = 3;

decltype((f.a)) e1 = x;          // int&
decltype((f.b)) e2 = y;          // int volatile&
decltype((f.c)) e3 = z;          // int const&

decltype((cf.a)) e4 = x;         // int const&
decltype((cf.b)) e5 = y;         // int const volatile&
decltype((cf.c)) e6 = z;         // int const&

decltype((pf->a)) e7 = x;        // int volatile&
decltype((pf->b)) e8 = y;        // int volatile&
decltype((pf->c)) e9 = z;        // int const volatile&

decltype((foo{}.a)) e10 = 0;   // int&&
decltype((foo{}.b)) e11 = 0;   // int volatile&&
decltype((foo{}.c)) e12 = 0;   // int const&&
```

Here, all the expressions used with decltype for declaring variables e1 to e9 are lvalues, so the deduced type is an lvalue reference. On the other hand, the expression used to declare the variables e10, e11, and e12 is an rvalue; therefore, the deduced type is an rvalue reference. Furthermore, cf is a constant object and foo::a has the type int. Therefore, the result type is const int&. Similarly, foo::b has the type volatile int; therefore, the result type is const volatile int&. These are just a few examples from this snippet, but the others follow the same rules for deduction.

Because `decltype` is a type specifier, the redundant `const` and `volatile` qualifiers and reference specifiers are ignored. This is demonstrated with the following example:

```
int a = 0;
int& ra = a;
int const c = 42;
int volatile d = 99;

decltype(ra)& e1 = a;        // int&
decltype(c) const e2 = 1;    // int const
decltype(d) volatile e3 = 1; // int volatile
```

So far in this section, we have learned how the `decltype` specifier works. However, its real purpose is to be used in templates, where the return value of a function depends on its template arguments and is not known before instantiation. To understand this scenario, let's start with the following example of a function template that returns the minimum of two values:

```
template <typename T>
T minimum(T&& a, T&& b)
{
    return a < b ? a : b;
}
```

We can use this as follows:

```
auto m1 = minimum(1, 5);          // OK
auto m2 = minimum(18.49, 9.99);// OK
auto m3 = minimum(1, 9.99);
                    // error, arguments of different type
```

The first two calls are both correct, as the supplied arguments are of the same type. The third call, however, will produce a compiler error, because the arguments have different types. For this to work, we need to cast the integer value to a `double`. However, there is an alternative: we could write a function template that takes two arguments of potentially different types and returns the minimum of the two. This can look as follows:

```
template <typename T, typename U>
??? minimum(T&& a, U&& b)
{
```

```
    return a < b ? a : b;
}
```

The question is, what is the return type of this function? This can be implemented differently, depending on the standard version you are using.

In C++11, we can use the `auto` specifier with a trailing return type, where we use the `decltype` specifier to deduce the return type from an expression. This would look as follows:

```
template <typename T, typename U>
auto minimum(T&& a, U&& b) -> decltype(a < b ? a : b)
{
    return a < b ? a : b;
}
```

This syntax can be simplified if you're using C++14 or a newer version of the standard. The trailing return type is no longer necessary. You can write the same function as follows:

```
template <typename T, typename U>
decltype(auto) minimum(T&& a, U&& b)
{
    return a < b ? a : b;
}
```

It is possible to simplify this further and simply use `auto` for the return type, shown as follows:

```
template <typename T, typename U>
auto minimum(T&& a, U&& b)
{
    return a < b ? a : b;
}
```

Although `decltype(auto)` and `auto` have the same effect in this example, this is not always the case. Let's consider the following example where we have a function returning a reference, and another function that calls it perfectly forwarding the argument:

```
template <typename T>
T const& func(T const& ref)
{
```

```
    return ref;
}


template <typename T>
auto func_caller(T&& ref)
{
    return func(std::forward<T>(ref));
}


int a = 42;
decltype(func(a))         r1 = func(a);         // int const&
decltype(func_caller(a)) r2 = func_caller(a); // int
```

The function func returns a reference, and func_caller is supposed to do a perfect forwarding to this function. By using auto for the return type, it is inferred as int in the preceding snippet (see variable r2). In order to do a perfect forwarding of the return type, we must use decltype(auto) for it, as shown next:

```
template <typename T>
decltype(auto) func_caller(T&& ref)
{
    return func(std::forward<T>(ref));
}


int a = 42;
decltype(func(a))         r1 = func(a);         // int const&
decltype(func_caller(a)) r2 = func_caller(a); // int const&
```

This time, the result is as intended, and the type of both r1 and r2 in this snippet is int const&.

As we have seen in this section, decltype is a type specifier used to deduce the type of an expression. It can be used in different contexts, but its purpose is for templates to determine the return type of a function and to ensure the perfect forwarding of it. Another feature that comes together with decltype is std::declval, which we will look at in the following section.

The std::declval type operator

The std::declval is a utility type operation function, available in the <utility> header. It's in the same category as functions such as std::move and std::forward that we have already seen. What it does is very simple: it adds an rvalue reference to its type template argument. The declaration of this function looks as follows:

```
template<class T>
typename std::add_rvalue_reference<T>::type declval() noexcept;
```

This function has no definition and therefore it cannot be called directly. It can only be used in **unevaluated contexts** – decltype, sizeof, typeid, and noexcept. These are compile-time-only contexts that are not evaluated during runtime. The purpose of std::declval is to aid with dependent type evaluation for types that do not have a default constructor or have one but it cannot be accessed because it's private or protected.

To understand how this works, let's consider a class template that does the composition of two values of different types, and we want to create a type alias for the result of applying the plus operator to two values of these types. How could such a type alias be defined? Let's start with the following form:

```
template <typename T, typename U>
struct composition
{
    using result_type = decltype(???);
};
```

We can use the decltype specifier but we need to provide an expression. We cannot say decltype(T + U) because these are types, not values. We could invoke the default constructor and, therefore, use the expression decltype(T{} + U{}). This can work fine for built-in types such as int and double, as shown in the following snippet:

```
static_assert(
  std::is_same_v<double,
                      composition<int, double>::result_type>);
```

It can also work for types that have an (accessible) default constructor. But it cannot work for types that don't have a default constructor. The following type wrapper is such an example:

```
struct wrapper
{
```

```
    wrapper(int const v) : value(v){}

    int value;

    friend wrapper operator+(int const a, wrapper const& w)
    {
        return wrapper(a + w.value);
    }

    friend wrapper operator+(wrapper const& w, int const a)
    {
        return wrapper(a + w.value);
    }
};


// error, no appropriate default constructor available
static_assert(
  std::is_same_v<wrapper,
                 composition<int,wrapper>::result_type>);
```

The solution here is to use `std::declval()`. The implementation of the class template composition would change as follows:

```
template <typename T, typename U>
struct composition
{
    using result_type = decltype(std::declval<T>() +
                                 std::declval<U>());
};
```

With this change, both the static asserts previously shown compile without any error. This function avoids the need to use particular values to determine the type of an expression. It produces a value of a type T without involving a default constructor. The reason it returns an rvalue reference is to enable us to work with types that cannot be returned from a function, such as arrays and abstract types.

The definition of the `wrapper` class earlier contained two friend operators. Friendship, when templates are involved, has some particularities. We will discuss this in the next section.

Understanding friendship in templates

When you define a class, you can restrict access to its member data and member functions with the `protected` and `private` access specifiers. If a member is private, it can only be accessed within the class. If a member is protected, it can be accessed from derived classes with public or protected access. However, a class can grant access to its private or protected members to other functions or classes with the help of the `friend` keyword. These functions or classes, to which special access has been granted, are called **friends**. Let's take a look at a simple example:

```
struct wrapper
{
    wrapper(int const v) :value(v) {}
private:
    int value;

    friend void print(wrapper const & w);
};

void print(wrapper const& w)
{ std::cout << w.value << '\n'; }

wrapper w{ 42 };
print(w);
```

The `wrapper` class has a private data member called `value`. There is a free function called `print` that takes an argument of the type `wrapper` and prints the wrapped value to the console. However, in order to be able to access it, the function is declared a friend of the `wrapper` class.

We will not focus on the way friendship works for non-templates. You should be familiar with this feature to proceed to discuss it in the context of templates. When it comes to templates, things get a bit complicated. We will look into this with the help of several examples. Let's start with the following:

```
struct wrapper
{
    wrapper(int const v) :value(v) {}
private:
```

```cpp
    int value;

    template <typename T>
    friend void print(wrapper const&);

    template <typename T>
    friend struct printer;
};

template <typename T>
void print(wrapper const& w)
{ std::cout << w.value << '\n'; }

template <typename T>
struct printer
{
    void operator()(wrapper const& w)
    { std::cout << w.value << '\n'; }
};

wrapper w{ 42 };
print<int>(w);
print<char>(w);
printer<int>()(w);
printer<double>()(w);
```

The print function is now a function template. It has a type template parameter, but that's not really used anywhere. That may look a bit odd, but it's a valid code, and we need to invoke it by specifying the template argument. However, it helps us make a point: any template instantiation of print, regardless of the template argument, can access the private members of the wrapper class. Notice the syntax used to declare it as a friend of the wrapper class: it uses the template syntax. The same applies to the class template printer. It's declared as a friend of the wrapper class and any template instantiation, regardless of the template argument, can access its private parts.

What if we wanted to restrict access to only some instances of these templates? Such as only the specializations for the `int` type? Then, we can declare these specializations as friends, as shown here:

```
struct wrapper;

template <typename T>
void print(wrapper const& w);

template <typename T>
struct printer;

struct wrapper
{
    wrapper(int const v) :value(v) {}
private:
    int value;

    friend void print<int>(wrapper const&);
    friend struct printer<int>;
};

template <typename T>
void print(wrapper const& w)
{ std::cout << w.value << '\n'; /* error */ }

template <>
void print<int>(wrapper const& w)
{ std::cout << w.value << '\n'; }

template <typename T>
struct printer
{
    void operator()(wrapper const& w)
    { std::cout << w.value << '\n'; /* error*/ }
```

```
};

template <>
struct printer<int>
{
    void operator()(wrapper const& w)
    { std::cout << w.value << '\n'; }
};

wrapper w{ 43 };
print<int>(w);
print<char>(w);
printer<int>()(w);
printer<double>()(w);
```

In this snippet, the `wrapper` class is the same as previously. For both the `print` function template and the `printer` class template, we have a primary template and a full specialization for the `int` type. Only the `int` instantiations are declared friends of the `wrapper` class. Attempting to access the private parts of the `wrapper` class in the primary templates generates compiler errors.

In these examples, the class that granted friendship to its private parts was a non-template class. But class templates can also declare friends. Let's see how it works in this case. We will start with the case of a class template and a non-template function:

```
template <typename T>
struct wrapper
{
    wrapper(T const v) :value(v) {}
private:
    T value;

    friend void print(wrapper<int> const&);
};

void print(wrapper<int> const& w)
{ std::cout << w.value << '\n'; }
```

```
void print(wrapper<char> const& w)
{ std::cout << w.value << '\n'; /* error */ }
```

In this implementation, the `wrapper` class template declares the overload of print that takes a `wrapper<int>` as a parameter as being a friend. Therefore, in this overloaded function, we can access the private data member `value`, but not in any other overload. A similar case occurs when the friend function or class is a template and we want only one specialization to access the private parts. Let's see the following snippet:

```
template <typename T>
struct printer;

template <typename T>
struct wrapper
{
    wrapper(T const v) :value(v) {}
private:
    T value;

    friend void print<int>(wrapper<int> const&);
    friend struct printer<int>;
};

template <typename T>
void print(wrapper<T> const& w)
{ std::cout << w.value << '\n'; /* error */ }

template<>
void print(wrapper<int> const& w)
{ std::cout << w.value << '\n'; }

template <typename T>
struct printer
{
    void operator()(wrapper<T> const& w)
    { std::cout << w.value << '\n'; /* error */ }
```

```
};

template <>
struct printer<int>
{
    void operator()(wrapper<int> const& w)
    { std::cout << w.value << '\n'; }
};
```

This implementation of the `wrapper` class template grants friendship to the `int` specializations of the `print` function template and `printer` class template. The attempt to access the private data member `value` in the primary templates (or any other specialization) would generate a compiler error.

If the intention is that the `wrapper` class template gives friend access to any instantiation of the `print` function template or `printer` class template, then the syntax to do so is as follows:

```
template <typename T>
struct printer;

template <typename T>
struct wrapper
{
    wrapper(T const v) :value(v) {}
private:
    T value;

    template <typename U>
    friend void print(wrapper<U> const&);

    template <typename U>
    friend struct printer;
};

template <typename T>
void print(wrapper<T> const& w)
```

```
{  std::cout << w.value << '\n'; }

template <typename T>
struct printer
{
    void operator()(wrapper<T> const& w)
    {  std::cout << w.value << '\n';  }
};
```

Notice that in declaring the friends, the syntax is `template <typename U>` and not `template <typename T>`. The name of the template parameter, U, can be anything except for T. That would shadow the name of the template parameter of the `wrapper` class template and that is an error. Keep in mind though that with this syntax, any specialization of `print` or `printer` has access to the private members of any specialization of the `wrapper` class template. If you want that only the specializations of the friends that meet the template argument of the wrapper class have access to its private parts, then you must use the following syntax:

```
template <typename T>
struct wrapper
{
    wrapper(T const v)  :value(v)  {}
private:
    T value;

    friend void print<T>(wrapper<T> const&);
    friend struct printer<T>;
};
```

This is similar to what we have seen previously when access was granted only to the int specializations, except that now it's for any specialization that matches T.

Apart from these cases, it's also possible for a class template to grant friendship to a type template parameter. This is demonstrated with the following example:

```
template <typename T>
struct connection
{
    connection(std::string const& host, int const port)
        :ConnectionString(host + ":" + std::to_string(port))
```

```
    {}
private:
    std::string ConnectionString;
    friend T;
};


struct executor
{
    void run()
    {
        connection<executor> c("localhost", 1234);

        std::cout << c.ConnectionString << '\n';
    }
};
```

The `connection` class template has a private data member called `ConnectionString`. The type template parameter `T` is a friend of the class. The `executor` class uses the instantiation `connection<executor>`, which means the `executor` type is the template argument and benefits from the friendship with the `connection` class so that it can access the private data member `ConnectionString`.

As can be seen from all these examples, friendship with templates is slightly different than friendship among non-template entities. Remember that friends have access to all the non-public members of a class. Therefore, friendship should be granted with care. On the other hand, if you need to grant access to some private members but not all, this is possible with the help of the **client-attorney pattern**. This pattern allows you to control the granularity of access to the private parts of a class. You can learn more about this pattern at this URL: `https://en.wikibooks.org/wiki/More_C%2B%2B_Idioms/Friendship_and_the_Attorney-Client`.

Summary

In this chapter, we went through a series of advanced topics. We started with name binding and dependent names and learned how to use the `typename` and the `template` keywords to tell the compiler what kind of dependent names we are referring to. Then, we learned about recursive templates and how to implement compile-time versions, using different approaches, for a recursive function.

We also learned about argument deduction for both function templates and class templates and how to help the compiler to do the latter with the help of user-defined deduction guides. An important topic covered in this chapter was the forwarding references and how they help us to implement perfect forwarding. Toward the end of the chapter, we learned about the `decltype` type specifier, the `std::declvalue` type utility, and, lastly, how friendship works in the context of class templates.

In the next chapter, we begin utilizing the knowledge accumulated so far about templates to do template metaprogramming, which is, basically, writing code that is evaluated at compile-time.

Questions

1. When is name lookup performed?
2. What are deduction guides?
3. What are forwarding references?
4. What does `decltype` do?
5. What does `std::declval` do?

Further readings

*   *Dependent name lookup for C++ templates* – Eli Bendersky, `https://eli.thegreenplace.net/2012/02/06/dependent-name-lookup-for-c-templates`
*   *Universal References in C++11* – Scott Meyers, `https://isocpp.org/blog/2012/11/universal-references-in-c11-scott-meyers`
*   *C++ Rvalue References Explained* – Thomas Becker, `http://thbecker.net/articles/rvalue_references/section_01.html`
*   *Universal vs Forwarding References in C++* – Petr Zemek, `https://blog.petrzemek.net/2016/09/17/universal-vs-forwarding-references-in-cpp/`

5

Type Traits and Conditional Compilation

Type traits are an important metaprogramming technique that enables us to inspect properties of types or to perform transformations of types at compile-time. Type traits are themselves templates and you can see them as meta-types. Knowing information such as the nature of a type, its supported operations, and its various properties is key for performing conditional compilation of templated code. It is also very useful when writing a library of templates.

In this chapter, you will learn the following:

- Understanding and defining type traits
- Understanding SFINAE and its purpose
- Enabling SFINAE with the `enable_if` type trait
- Using `constexpr if`
- Exploring the standard type traits
- Seeing real-world examples of using type traits

By the end of the chapter, you will have a good understanding of what type traits are, how they are useful, and what type traits are available in the **C++** standard library.

We will start the chapter by looking at what type traits are and how they help us.

Understanding and defining type traits

In a nutshell, **type traits** are small class templates that contain a constant value whose value represents the answer to a question we ask about a type. An example of such a question is: is this type a floating-point type? The technique for building type traits that provide such information about types relies on template specialization: we define a primary template as well as one or more specializations.

Let's see how we can build a type trait that tells us, at compile-time, whether a type is a floating-point type:

```cpp
template <typename T>
struct is_floating_point
{
    static const bool value = false;
};


template <>
struct is_floating_point<float>
{
    static const bool value = true;
};


template <>
struct is_floating_point<double>
{
    static const bool value = true;
};


template <>
struct is_floating_point<long double>
{
    static const bool value = true;
};
```

There are two things to notice here:

- We have defined a primary template as well as several full specializations, one for each type that is a floating-point type.

- The primary template has a `static const` Boolean member initialized with the `false` value; the full specializations set the value of this member to `true`.

There is nothing more to building a type trait than this. `is_floating_point<T>` is a type trait that tells us whether a type is a floating-point type or not. We can use it as follows:

```
int main()
{
    static_assert(is_floating_point<float>::value);
    static_assert(is_floating_point<double>::value);
    static_assert(is_floating_point<long double>::value);
    static_assert(!is_floating_point<int>::value);
    static_assert(!is_floating_point<bool>::value);
}
```

This proves that we have built the type trait correctly. But it does not show a real use-case scenario. For this type trait to be really useful, we need to use it at compile-time to do something with the information it provides.

Let's suppose we want to build a function that does something with a floating-point value. There are multiple floating-point types, such as `float`, `double`, and `long double`. For us to avoid writing multiple implementations, we would build this as a template function. However, that means we could actually pass other types as template arguments, so we need a way to prevent that. A simple solution is to use the `static_assert()` statement we saw earlier and produce an error should the user supply a value that is not a floating-point value. This can look as follows:

```
template <typename T>
void process_real_number(T const value)
{
    static_assert(is_floating_point<T>::value);

    std::cout << "processing a real number: " << value
              << '\n';
```

```
}

int main()
{
    process_real_number(42.0);
    process_real_number(42); // error:
                             // static assertion failed
}
```

This is a really simple example but it demonstrates the use of type traits to do conditional compilation. There are other approaches than using static_assert() and we will explore them throughout this chapter. For the time being, let's look at a second example.

Suppose we have classes that define operations for writing to an output stream. This is basically a form of serialization. However, some support this with an overloaded operator<<, others with the help of a member function called write. The following listing shows two such classes:

```
struct widget
{
    int         id;
    std::string name;

    std::ostream& write(std::ostream& os) const
    {
        os << id << ',' << name << '\n';
        return os;
    }
};

struct gadget
{
    int         id;
    std::string name;

    friend std::ostream& operator <<(std::ostream& os,
                                      gadget const& o);
```

```
};

std::ostream& operator <<(std::ostream& os,
                          gadget const& o)
{
    os << o.id << ',' << o.name << '\n';
    return os;
}
```

In this example, the `widget` class contains a member function, `write`. However, for the `gadget` class, the stream operator, `<<`, is overloaded for the same purpose. We can write the following code using these classes:

```
widget w{ 1, "one" };
w.write(std::cout);

gadget g{ 2, "two" };
std::cout << g;
```

However, our goal would be to define a function template that enables us to treat them the same way. In other words, instead of using either `write` or the `<<` operator, we should be able to write the following:

```
serialize(std::cout, w);
serialize(std::cout, g);
```

This brings up some questions. First, how would such a function template look, and second, how can we know whether a type provides a `write` method or has the `<<` operator overloaded? The answer to the second question is type traits. We can build a type trait to help us answer this latter question at compile-time. This is how such a type trait may look:

```
template <typename T>
struct uses_write
{
    static constexpr bool value = false;
};

template <>
struct uses_write<widget>
{
```

```
    static constexpr bool value = true;
};
```

This is very similar to the type trait we defined previously. `uses_write` tells us whether a type defines the `write` member function. The primary template sets the data member called `value` to `false`, but the full specialization for the `widget` class sets it to `true`. In order to avoid the verbose syntax `uses_write<T>::value`, we can also define a variable template, reducing the syntax to the form `uses_write_v<T>`. This variable template will look as follows:

```
template <typename T>
inline constexpr bool uses_write_v = uses_write<T>::value;
```

To make the exercise simple, we'll assume that the types that don't provide a `write` member function overload the output stream operator. In practice, this is would not be the case, but for the sake of simplicity, we will build on this assumption.

The next step in defining the function template `serialize` that provides a uniform API for serializing all classes is to define more class templates. However, these would follow the same path – a primary template that provides one form of serialization and a full specialization that provides a different form. Here is the code for it:

```
template <bool>
struct serializer
{
    template <typename T>
    static void serialize(std::ostream& os, T const& value)
    {
        os << value;
    }
};

template<>
struct serializer<true>
{
    template <typename T>
    static void serialize(std::ostream& os, T const& value)
    {
        value.write(os);
```

```
    }
};
```

The `serializer` class template has a single template parameter, which is a non-type template parameter. It is also an anonymous template parameter because we don't use it anywhere in the implementation. This class template contains a single member function. It is actually a member function template with a single type template parameter. This parameter defines the type of value we would serialize. The primary template uses the `<<` operator to output the value to the provided stream. On the other hand, the full specialization of the `serializer` class template uses the member function `write` to do the same. Notice that we fully specialize the `serializer` class template and not the `serialize` member function template.

The only thing left now is to implement the desired free function `serialize`. Its implementation will be based on the `serializer<T>::serialize` function. Let's see how:

```
template <typename T>
void serialize(std::ostream& os, T const& value)
{
    serializer<uses_write_v<T>>::serialize(os, value);
}
```

The signature of this function template is the same as the one of the `serialize` member function from the `serializer` class template. The selection between the primary template and the full specialization is done with the help of the variable template `uses_write_v`, which provides a convenient way to access the value data member of the `uses_write` type trait.

In these examples, we have seen how to implement type traits and use the information they provide at compile-time to either impose restrictions on types or select between one implementation or the other. A similar purpose has another metaprogramming technique called **SFINAE**, which we will cover next.

Exploring SFINAE and its purpose

When we write templates, we sometimes need to restrict the template arguments. For instance, we have a function template that should work for any numeric type, therefore integral and floating-point, but should not work with anything else. Or we may have a class template that should only accept trivial types for an argument.

There are also cases when we may have overloaded function templates that should each work with some types only. For instance, one overload should work for integral types and the other for floating-point types only. There are different ways to achieve this goal and we will explore them in this chapter and the next.

Type traits, however, are involved in one way or another in all of them. The first one that will be discussed in this chapter is a feature called SFINAE. Another approach, superior to SFINAE, is represented by concepts, which will be discussed in the next chapter.

SFINAE stands for **Substitution Failure Is Not An Error**. When the compiler encounters the use of a function template, it substitutes the arguments in order to instantiate the template. If an error occurs at this point, it is not regarded as ill-informed code, only as a deduction failure. The function is removed from the overload set instead of causing an error. Only if there is no match in the overload set does an error occur.

It's difficult to really understand SFINAE without concrete examples. Therefore, we will go through several examples to explain the concept.

Every standard container, such as `std::vector`, `std::array`, and `std::map`, not only has iterators that enable us to access its elements but also modify the container (such as inserting after the element pointed by an iterator). Therefore, these containers have member functions to return iterators to the first and the one-past-last elements of the container. These methods are called `begin` and `end`.

There are other methods such as `cbegin` and `cend`, `rbegin` and `rend`, and `crbegin` and `crend` but these are beyond the purpose of this topic. In C++11, there are also free functions, `std:begin` and `std::end`, that do the same. However, these work not just with standard containers but also with arrays. One benefit of these is enabling range-based `for` loops for arrays. The question is how this non-member function could be implemented to work with both containers and arrays? Certainly, we need two overloads of a function template. A possible implementation is the following:

```
template <typename T>
auto begin(T& c) { return c.begin(); }    // [1]


template <typename T, size_t N>
T* begin(T(&arr)[N]) {return arr; }        // [2]
```

The first overload calls the member function `begin` and returns the value. Therefore, this overload is restricted to types that have a member function `begin`; otherwise, a compiler error would occur. The second overload simply returns a pointer to the first element of the array. This is restricted to array types; anything else would produce a compiler error. We can use these overloads as follows:

```
std::array<int, 5> arr1{ 1,2,3,4,5 };
std::cout << *begin(arr1) << '\n';          // [3] prints 1


int arr2[]{ 5,4,3,2,1 };
std::cout << *begin(arr2) << '\n';          // [4] prints 5
```

If you compile this piece of code, no error, not even a warning, occurs. The reason for that is SFINAE. When resolving the call to `begin(arr1)`, substituting `std::array<int, 5>` to the first overload (at `[1]`) succeeds, but the substitution for the second (at `[2]`) fails. Instead of issuing an error at this point, the compiler just ignores it, so it builds an overload set with a single instantiation and, therefore, it can successfully find a match for the invocation. Similarly, when resolving the call to `begin(arr2)`, the substitution of `int [5]` for the first overload fails and is ignored, but it succeeds for the second and is added to the overload set, eventually finding a good match for the invocation. Therefore, both calls can be successfully made. Should one of the two overloads not be present, either `begin(arr1)` or `begin(arr2)` would fail to match the function template and a compiler error would occur.

SFINAE only applies in the so-called **immediate context** of a function. The immediate context is basically the template declaration (including the template parameter list, the function return type, and the function parameter list). Therefore, it does not apply to the body of a function. Let's consider the following example:

```
template <typename T>
void increment(T& val) { val++; }


int a = 42;
increment(a);   // OK


std::string s{ "42" };
increment(s);   // error
```

There are no restrictions on the type `T` in the immediate context of the `increment` function template. However, in the body of the function, the parameter `val` is incremented with the post-fix `operator++`. That means, substituting for `T` any type for which the post-fix `operator++` is not implemented is a failure. However, this failure is an error and will not be ignored by the compiler.

The C++ standard (license usage link: `http://creativecommons.org/licenses/by-sa/3.0/`) defines the list of errors that are considered SFINAE errors (in paragraph *§13.10.2, Template argument deduction*, the **C++20** standard version). These SFINAE errors are the following attempts:

- Creating an array of `void`, an array of reference, an array of function, an array of negative size, an array of size zero, and an array of non-integral size
- Using a type that is not a class or enum on the left side of the scope resolution operator `::` (such as in `T::value_type` with `T` being a numeric type for instance)
- Creating a pointer to reference
- Creating a reference to `void`
- Creating a pointer to member of `T`, where `T` is not a class type
- Using a member of a type when the type does not contain that member
- Using a member of a type where a type is required but the member is not a type
- Using a member of a type where a template is required but the member is not a template
- Using a member of a type where a non-type is required but the member is not a non-type
- Creating a function type with a parameter of type `void`
- Creating a function type that returns an array type or another function type
- Performing an invalid conversion in a template argument expression or an expression used in a function declaration
- Supplying an invalid type to a non-type template parameter
- Instantiating a pack expansion containing multiple packs of different lengths

The last error in this list was introduced in C++11 together with variadic templates. The others were defined before C++11. We will not go on to exemplify all of these errors, but we can take a look at a couple more examples. The first concerns attempting to create an array of size zero. Let's say we want to have two function template overloads, one that handles arrays of even sizes and one that handles arrays of odd sizes. A solution to this is the following:

```
template <typename T, size_t N>
void handle(T(&arr)[N], char(*)[N % 2 == 0] = 0)
{
    std::cout << "handle even array\n";
}

template <typename T, size_t N>
void handle(T(&arr)[N], char(*)[N % 2 == 1] = 0)
{
    std::cout << "handle odd array\n";
}

int arr1[]{ 1,2,3,4,5 };
handle(arr1);

int arr2[]{ 1,2,3,4 };
handle(arr2);
```

The template arguments and the first function parameter are similar to what we saw with the `begin` overload for arrays. However, these overloads for `handle` have a second anonymous parameter with the default value `0`. The type of this parameter is a pointer to an array of type `char` and a size specified with the expressions `N%2==0` and `N%2==1`. For every possible array, one of these two is `true` and the other is `false`. Therefore, the second parameter is either `char(*)[1]` or `char(*)[0]`, the latter being an SFINAE error (an attempt to create an array of size zero). Therefore, we are able to call either one of the other overloads without generating compiler errors, thanks to SFINAE.

The last example that we will look at in this section will show SFINAE with an attempt to use a member of a class that does not exist. Let's start with the following snippet:

```
template <typename T>
struct foo
{
```

```
    using foo_type = T;
};


template <typename T>
struct bar
{
    using bar_type = T;
};


struct int_foo : foo<int> {};
struct int_bar : bar<int> {};
```

Here we have two classes, foo, which has a member type called foo_type, and bar, which has a member type called bar_type. There are also classes that derive from these two. The goal is to write two function templates, one that handles the foo hierarchy of classes, and one that handles the bar hierarchy of classes. A possible implementation is the following:

```
template <typename T>
decltype(typename T::foo_type(), void()) handle(T const& v)
{
    std::cout << "handle a foo\n";
}


template <typename T>
decltype(typename T::bar_type(), void()) handle(T const& v)
{
    std::cout << "handle a bar\n";
}
```

Both overloads have a single template parameter and a single function parameter of type T const&. They also return the same type, and that type is void. The expression decltype(typename T::foo_type(), void()) may need a little consideration to understand better. We discussed decltype in *Chapter 4, Advanced Template Concepts*. Remember that this is a type specifier that deduces the type of an expression. We use the comma operator, so the first argument is evaluated but then discarded, so decltype will only deduce the type from `void()`, and the deduced type is `void`. However, the arguments `typename T::foo_type()` and `typename T::bar_type()` do use an inner type, and this only exists either for `foo` or `bar`. This is where SFINAE manifests itself, as shown in the following snippet:

```
int_foo fi;
int_bar bi;
int x = 0;
handle(fi); // OK
handle(bi); // OK
handle(x);  // error
```

Calling `handle` with an `int_foo` value will match the first overload, while the second is discarded because of a substitution failure. Similarly, calling `handle` with an `int_bar` value will match the second overload, while the first is discarded because of a substitution failure. However, calling `handle` with an `int` will cause substitution failure for both overloads so the final overload set for substituting `int` will be empty, which means there is no match for the call. Therefore, a compiler error occurs.

SFINAE is not the best way to achieve conditional compilation. However, in modern C++ it's probably best used together with a type trait called `enable_if`. This is what we will discuss next.

Enabling SFINAE with the enable_if type trait

The C++ standard library is a family of sub-libraries. One of these is the type support library. This library defines types such as `std::size_t`, `std::nullptr_t`, and `std::byte`, run-time type identification support with classes such as `std::type_info`, as well as a collection of type traits. There are two categories of type traits:

- Type traits that enable us to query properties of types at compile-time.

- Type traits that enable us to perform type transformations at compile-time (such as adding or removing the `const` qualifier, or adding or removing pointer or reference from a type). These type traits are also called **metafunctions**.

One type trait from the second category is `std::enable_if`. This is used to enable SFINAE and remove candidates from a function's overload set. A possible implementation is the following:

```
template<bool B, typename T = void>
struct enable_if {};


template<typename T>
struct enable_if<true, T> { using type = T; };
```

There is a primary template, with two template parameters, a Boolean non-type template, and a type parameter with `void` as the default argument. This primary template is an empty class. There is also a partial specialization for the `true` value of the non-type template parameter. This, however, defines a member type simply called `type`, which is an alias template for the template parameter `T`.

The `enable_if` metafunction is intended to be used with a Boolean expression. When this Boolean expression is evaluated as `true`, it defines a member type called `type`. If the Boolean expression is `false`, this member type is not defined. Let's see how it works.

Remember the example from the *Understanding and defining type traits* section at the beginning of the chapter, where we had classes that provided a `write` method to write their content to an output stream, and classes for which the `operator<<` was overloaded for the same purpose? In that section, we defined a type trait called `uses_write` and wrote a `serialize` function template that allowed us to serialize, in a uniform manner, both types of objects (`widget` and `gadget`). However, the implementation was rather complex. With `enable_if`, we can implement that function in a simple manner. A possible implementation is shown in the next snippet:

```
template <typename T,
          typename std::enable_if<
              uses_write_v<T>>::type* = nullptr>
void serialize(std::ostream& os, T const& value)
{
    value.write(os);
}


template <typename T,
          typename std::enable_if<
              !uses_write_v<T>>::type*=nullptr>
```

```
void serialize(std::ostream& os, T const& value)
{
    os << value;
}
```

There are two overloaded function templates in this implementation. They both have two template parameters. The first parameter is a type template parameter, called `T`. The second is an anonymous non-type template parameter of a pointer type that also has the default value `nullptr`. We use `enable_if` to define the member called `type` only if the `uses_write_v` variable evaluates to `true`. Therefore, for classes that have the member function `write`, the substitution succeeds for the first overload but fails for the second overload, because `typename * = nullptr` is not a valid parameter. For classes for which the `operator<<` is overloaded, we have the opposite situation.

The `enable_if` metafunction can be used in several scenarios:

- To define a template parameter that has a default argument, which we saw earlier
- To define a function parameter that has a default argument
- To specify the return type of a function

For this reason, I mentioned earlier that the provided implementation of the `serialize` overloads is just one of the possibilities. A similar one that uses `enable_if` to define a function parameter with a default argument is shown next:

```
template <typename T>
void serialize(
    std::ostream& os, T const& value,
    typename std::enable_if<
               uses_write_v<T>>::type* = nullptr)
{
    value.write(os);
}

template <typename T>
void serialize(
    std::ostream& os, T const& value,
    typename std::enable_if<
               !uses_write_v<T>>::type* = nullptr)
{
```

```
    os << value;
}
```

You will notice here that we basically moved the parameter from the template parameter list to the function parameter list. There is no other change, and the usage is the same, such as follows:

```
widget w{ 1, "one" };
gadget g{ 2, "two" };

serialize(std::cout, w);
serialize(std::cout, g);
```

The third alternative is to use enable_if to wrap the return type of the function. This implementation is only slightly different (the default argument does not make sense for a return type). Here is how it looks:

```
template <typename T>
typename std::enable_if<uses_write_v<T>>::type serialize(
    std::ostream& os, T const& value)
{
    value.write(os);
}


template <typename T>
typename std::enable_if<!uses_write_v<T>>::type serialize(
    std::ostream& os, T const& value)
{
    os << value;
}
```

The return type, in this implementation, is defined if uses_write_v<T> is true. Otherwise, a substitution failure occurs and SFINAE takes place.

Although in all these examples, the `enable_if` type trait was used to enable SFINAE during the overload resolution for function templates, this type trait can also be used to restrict instantiations of class templates. In the following example, we have a class called `integral_wrapper` that is supposed to be instantiated only with integral types, and a class called `floating_wrapper` that is supposed to be instantiated only with floating-point types:

```
template <
    typename T,
    typename=typenamestd::enable_if_t<
                        std::is_integral_v<T>>>
struct integral_wrapper
{
    T value;
};


template <
    typename T,
    typename=typename std::enable_if_t<
                        std::is_floating_point_v<T>>>
struct floating_wrapper
{
    T value;
};
```

Both these class templates have two type template parameters. The first one is called `T`, but the second one is anonymous and has a default argument. The value of this argument is defined or not with the help of the `enable_if` type trait, based on the value of a Boolean expression.

In this implementation, we can see:

- An alias template called `std::enable_if_t`, which is a convenient way to access the `std::enable_if<B, T>::type` member type. This is defined as follows:

```
template <bool B, typename T = void>
using enable_if_t = typename enable_if<B,T>::type;
```

- Two variable templates, std::is_integral_v and std::is_floating_
  point_v, which are convenient ways to access the data members, std::is_
  integral<T>::value and std::is_floating_point<T>::value. The
  std::is_integral and std::is_floating_point classes are standard
  type traits that check whether a type is an integral type or a floating-point
  type respectively.

The two wrapper class templates shown previously can be used as follows:

```
integral_wrapper w1{ 42 };    // OK
integral_wrapper w2{ 42.0 };  // error
integral_wrapper w3{ "42" };  // error


floating_wrapper w4{ 42 };    // error
floating_wrapper w5{ 42.0 };  // OK
floating_wrapper w6{ "42" };  // error
```

Only two of these instantiations work, w1, because integral_wrapper is instantiated
with the int type, and w5, because floating_wrapper is instantiated with the
double type. All the others generate compiler errors.

It should be pointed out that this code samples only work with the provided definitions of
integral_wrapper and floating_wrapper in C++20. For previous versions of the
standard, even the definitions of w1 and w5 would generate compiler errors because the
compiler wasn't able to deduce the template arguments. In order to make them work, we'd
have to change the class templates to include a constructor, as follows:

```
template <
    typename T,
    typename=typenamestd::enable_if_t<
                        std::is_integral_v<T>>>
struct integral_wrapper
{
    T value;

    integral_wrapper(T v) : value(v) {}
};

template <
```

```
    typename T,
    typename=typename std::enable_if_t<
                         std::is_floating_point_v<T>>>
struct floating_wrapper
{
    T value;

    floating_wrapper(T v) : value(v) {}
};
```

Although `enable_if` helps achieve SFINAE with simpler and more readable code, it's still rather complicated. Fortunately, in **C++17** there is a better alternative with `constexpr if` statements. Let's explore this alternative next.

Using constexpr if

A C++17 feature makes SFINAE much easier. It's called `constexpr if` and it's a compile-time version of the `if` statement. It helps replace complex template code with simpler versions. Let's start by looking at a C++17 implementation of the `serialize` function that can uniformly serialize both widgets and gadgets:

```
template <typename T>
void serialize(std::ostream& os, T const& value)
{
    if constexpr (uses_write_v<T>)
        value.write(os);
    else
        os << value;
}
```

The syntax for `constexpr if` is `if constexpr (condition)`. The condition must be a compile-time expression. There is no short-circuit logic performed when evaluating the expression. This means that if the expression has the form `a && b` or `a || b`, then both a and b must be well-formed.

constexpr if enables us to discard a branch, at compile-time, based on the value of the expression. In our example, when the uses_write_v variable is true, the else branch is discarded, and the body of the first branch is retained. Otherwise, the opposite occurs. Therefore, we end up with the following specializations for the widget and gadget classes:

```
template<>
void serialize<widget>(std::ostream & os,
                       widget const & value)
{
    if constexpr (true)
    {
        value.write(os);
    }
}

template<>
void serialize<gadget>(std::ostream & os,
                       gadget const & value)
{
    if constexpr (false)
    {
    }
    else
    {
        os << value;
    }
}
```

Of course, this code is likely to be further simplified by the compiler. Therefore, eventually, these specializations would simply look like the following:

```
template<>
void serialize<widget>(std::ostream & os,
                       widget const & value)
{
    value.write(os);
```

```
}

template<>
void serialize<gadget>(std::ostream & os,
                       gadget const & value)
{
    os << value;
}
```

The end result is the same as the one we achieved with SFINAE and `enable_if`, but the actual code we wrote here was simpler and easier to understand.

`constexpr if` is a great tool for simplifying code and we actually saw it earlier in *Chapter 3, Variadic Templates*, in the *Parameter packs* paragraph, when we implemented a function called `sum`. This is shown again here:

```
template <typename T, typename... Args>
T sum(T a, Args... args)
{
    if constexpr (sizeof...(args) == 0)
        return a;
    else
        return a + sum(args...);
}
```

In this example, `constexpr if` helps us to avoid having two overloads, one for the general case and one for ending the recursion. Another example presented already in this book where `constexpr if` can simplify the implementation is the `factorial` function template from *Chapter 4, Advanced Template Concepts*, in the *Exploring template recursion* section. That function looked as follows:

```
template <unsigned int n>
constexpr unsigned int factorial()
{
    return n * factorial<n - 1>();
}

template<>
constexpr unsigned int factorial<1>() { return 1; }
```

```
template<>
constexpr unsigned int factorial<0>() { return 1; }
```

With `constexpr if`, we can replace all this with a single template and let the compiler take care of providing the right specializations. The C++17 version of this function may look as follows:

```
template <unsigned int n>
constexpr unsigned int factorial()
{
    if constexpr (n > 1)
        return n * factorial<n - 1>();
    else
        return 1;
}
```

The `constexpr if` statements can be useful in many situations. The last example presented in this section is a function template called `are_equal`, which determines whether the two supplied arguments are equal or not. Typically, you'd think that using `operator==` should be enough to determine whether two values are equal or not. That is true in most cases, except for floating-point values. Because only some of the floating-point numbers can be stored without a precision loss (numbers like 1, 1.25, 1.5, and anything else where the fractional part is an exact sum of inverse powers of 2) we need to take special care when comparing floating-point numbers. Usually, this is solved by ensuring that the difference between two floating-point values is less than some threshold. Therefore, a possible implementation for such a function could be as follows:

```
template <typename T>
bool are_equal(T const& a, T const& b)
{
    if constexpr (std::is_floating_point_v<T>)
        return std::abs(a - b) < 0.001;
    else
        return a == b;
}
```

When the T type is a floating-point type, we compare the absolute value of the difference of the two numbers with the selected threshold. Otherwise, we fall back to using operator==. This enables us to use this function not just with arithmetic types, but also any other type for which the equality operator is overloaded.

```
are_equal(1, 1);                                       // OK
are_equal(1.999998, 1.999997);                         // OK
are_equal(std::string{ "1" }, std::string{ "1" }); // OK
are_equal(widget{ 1, "one" }, widget{ 1, "two" }); // error
```

We are able to call the are_equal function template with arguments of type int, double, and std::string. However, attempting to do the same with values of the widget type will trigger a compiler error, because the == operator is not overloaded for this type.

So far in this chapter, we have seen what type traits are as well as different ways to perform conditional compilation. We have also seen some of the type traits available in the standard library. In the second part of this chapter, we will explore what the standard has to offer with regard to type traits.

Exploring the standard type traits

The standard library features a series of type traits for querying properties of types as well as performing transformations on types. These type traits are available in the <type_traits> header as part of the type support library. There are several categories of type traits including the following:

- Querying the type category (primary or composite)
- Querying type properties
- Querying supported operations
- Querying type relationships
- Modifying cv-specifiers, references, pointers, or a sign
- Miscellaneous transformations

Although looking at every single type trait is beyond the scope of this book, we will explore all these categories to see what they contain. In the following subsections, we will list the type traits (or most of them) that make up each of these categories. These lists as well as detailed information about each type trait can be found in the C++ standard (see the *Further reading* section at the end of the chapter for a link to a freely available draft version) or on the `cppreference.com` website at `https://en.cppreference.com/w/cpp/header/type_traits` (license usage link: `http://creativecommons.org/licenses/by-sa/3.0/`).

We will start with the type traits for querying the type category.

Querying the type category

Throughout this book so far, we have used several type traits, such as `std::is_integral`, `std::is_floating_point`, and `std::is_arithmetic`. These are just some of the standard type traits used for querying primary and composite type categories. The following table lists the entire set of such type traits:

Name	Description
`is_void`	Checks whether a type is the `void` type.
`is_null_pointer`	Checks whether a type is the `std::nullptr_t` type.
`is_integral`	Checks whether a type is an integral type, including signed, unsigned, and cv-qualified variants. The integral types are: • `bool`, `char`, `char8_t` (since C++20), `char16_t`, `char32_t`, `wchar_t`, `short`, `int`, `long`, and `long long` • Any implementation-defined extended integer type
`is_floating_point`	Checks whether a type is a floating-point type, including cv-qualified variants. The possible types are `float`, `double`, and `long double`.
`is_array`	Checks whether a type is an array type.
`is_enum`	Checks whether a type is an enumeration type.
`is_union`	Checks whether a type is a union type.
`is_class`	Checks whether a type is a class type, but not a union type.
`is_function`	Checks whether a type is a function type. This excludes lambdas, classes with overloaded call operator, pointers to functions, and the `std::function` type.

Name	Description
`is_pointer`	Checks whether a type is a pointer to object or a pointer to function or a cv-qualified variant. This does not include pointer to member object or pointer to member function.
`is_member_pointer`	Checks whether a type is a pointer to a non-static member object or pointer to a non-static member function.
`is_member_object_pointer`	Checks whether a type is a non-static member object pointer.
`is_member_function_pointer`	Checks whether a type is a non-static member function pointer.
`is_lvalue_reference`	Checks whether a type is an lvalue reference type.
`is_rvalue_reference`	Checks whether a type is an rvalue reference type.
`is_reference`	Checks whether a type is a reference type. This can be either an lvalue or rvalue reference type.
`is_fundamental`	Checks whether a type is a fundamental type. Fundamental types are the arithmetic types, the `void` type, and the `std::nullptr_t` type.
`is_scalar`	Checks whether a type is a scalar type or a cv-qualifier version of one. Scalar types are: • The arithmetic types • Pointer types • Pointer to member types • Enumeration types • The `std::nullptr_t` type
`is_object`	Checks whether a type is an object type of a cv-qualifier version. An object type is a type that is not a function type, a reference type, or the `void` type.

Name	Description
`is_compound`	Checks whether a type is a compound type or any cv-qualified variant of one. Compound types are types that are not fundamental. These are: • Array types • Function types • Class types • Union types • Object pointer and function pointer types • Member object pointer and member function pointer types • Reference types • Enumeration types

Table 5.1

All these type traits are available in C++11. For each of them, starting with C++17, a variable template is available to simplify the access to the Boolean member called `value`. For a type trait with the name `is_abc`, a variable template with the name `is_abc_v` exists. This is true for all the type traits that have a Boolean member called `value`. The definition of these variables is very simple. The next snippet shows the definition for the `is_arithmentic_v` variable template:

```
template< class T >
inline constexpr bool is_arithmetic_v =
    is_arithmetic<T>::value;
```

Here is an example of using some of these type traits:

```
template <typename T>
std::string as_string(T value)
{
    if constexpr (std::is_null_pointer_v<T>)
        return "null";
    else if constexpr (std::is_arithmetic_v<T>)
        return std::to_string(value);
    else
        static_assert(always_false<T>);
```

```
    }

    std::cout << as_string(nullptr) << '\n'; // prints null
    std::cout << as_string(true) << '\n';    // prints 1
    std::cout << as_string('a') << '\n';     // prints a
    std::cout << as_string(42) << '\n';      // prints 42
    std::cout << as_string(42.0) << '\n';    // prints 42.000000
    std::cout << as_string("42") << '\n';    // error
```

The function template `as_string` returns a string containing the value pass as an argument. It works with arithmetic types only and with the `nullptr_t` for which it returns the value `"null"`.

You must have noticed the statement, `static_assert(always_false<T>)`, and wondering what this `always_false<T>` expression actually is. It is a variable template of the `bool` type that evaluates to `false`. Its definition is as simple as the following:

```
template<class T>
constexpr bool always_false = std::false_type::value;
```

This is needed because the statement, `static_assert(false)`, would make the program ill-formed. The reason for this is that its condition would not depend on a template argument but evaluate to `false`. When no valid specialization can be generated for a sub-statement of a `constexpr if` statement within a template, the program is ill-formed (and no diagnostic is required). To avoid this, the condition of the `static_assert` statement must depend on a template argument. With `static_assert(always_false<T>)`, the compiler does not know whether this would evaluate to `true` or `false` until the template is instantiated.

The next category of type traits we explore allows us to query properties of types.

Querying type properties

The type traits that enable us to query properties of types are the following:

Name	C++ version	Description
is_const	C++11	Checks whether a type is const-qualified (either `const` or `const volatile`).
is_volatile	C++11	Checks whether a type is volatile-qualified (either `volatile` or `const volatile`).

Name	C++ version	Description
is_trivial	C++11	Checks whether a type is a trivial type or a cv-qualified variant. The following are trivial types: • Scalar types or arrays of a scalar type • Trivially copyable classes with a trivial default constructor or arrays of such a class
is_trivially_copyable	C++11	Checks whether a type is trivially copyable. The following are trivially copyable types: • Scalar types or arrays of a scalar type • Trivially copyable classes or arrays of such a class
is_standard_layout	C++11	Checks whether a type is a standard layout type or a cv-qualified variant. The following are such types: • Scalar types or arrays of a scalar type • Standard-layout classes or arrays of such a class
is_empty	C++11	Checks whether a type is an empty type. An empty type is a class type (that is not a union) and has no non-static data members (except for bit fields of size 0), no virtual functions, no virtual base classes, and no non-empty base classes.
is_polymorphic	C++11	Checks whether a type is a polymorphic type. A polymorphic type is a class type (that is not a union) that inherits at least one virtual function.
is_abstract	C++11	Checks whether a type is an abstract type. An abstract type is a class type (that is not a union) that inherits at least one virtual pure function.
is_final	C++14	Checks whether a type is a class type declared with the final specifier.
is_aggregate	C++17	Checks whether a type is an aggregate type.
is_signed	C++11	Checks whether a type is a floating-point type or a signed integral type.

Name	C++ version	Description
`is_unsigned`	C++11	Checks whether a type is an unsigned integral type or the `bool` type.
`is_bounded_array`	C++20	Checks whether a type is an array type of a known bound (such as `int [5]`).
`is_unbounded_array`	C++20	Checks whether a type is an array type of an unknown bound (such as `int []`).
`is_scoped_enum`	C++23	Checks whether a type is a scoped enumeration type.
`has_unique_object_representation`	C++17	Checks whether a type is trivially copyable, and any two objects of this type having the same value also have the same object representation.

Table 5.2

Although most of these are probably straightforward to understand, there are two that seem the same at a first glance. These are `is_trivial` and `is_trivially_copyable`. These both are true for scalar types or arrays of scalar types. They also are true for classes that are trivially copyable or arrays of such classes but `is_trivial` is true only for copyable classes that have a trivial default constructor.

According to the paragraph *§11.4.4.1* in the C++ 20 standard, a default constructor is trivial if it is not user-provided, and the class has no virtual member functions, no virtual base classes, no non-static members with default initializers, every direct base of it has a trivial default constructor, and every non-static member of a class type also has a trivial default constructor. To understand this better, let's look at the following example:

```
struct foo
{
    int a;
};


struct bar
{
    int a = 0;
};


struct tar
{
    int a = 0;
```

```
    tar() : a(0) {}
};


std::cout << std::is_trivial_v<foo> << '\n'; // true
std::cout << std::is_trivial_v<bar> << '\n'; // false
std::cout << std::is_trivial_v<tar> << '\n'; // false


std::cout << std::is_trivially_copyable_v<foo>
          << '\n';                                    // true
std::cout << std::is_trivially_copyable_v<bar>
          << '\n';                                    // true
std::cout << std::is_trivially_copyable_v<tar>
          << '\n';                                    // true
```

In this example, there are three similar classes. All three of them, foo, bar, and tar, are trivially copyable. However, only the foo class is a trivial class, because it has a trivial default constructor. The bar class has a non-static member with a default initializer, and the tar class has a user-defined constructor, and this makes them non-trivial.

Apart from trivial copy-ability, there are other operations that we can query for with the help of other type traits. We will see these in the following section.

Querying supported operations

The following set of type traits helps us to query supported operations:

Name	Description
is_constructible is_trivially_constructible is_nothrow_constructible	Check whether a type has a constructor that can take specific arguments.
is_default_constructible is_trivially_default_constructible is_nothrow_default_constructible	Check whether a type has a default constructor.
is_copy_constructible is_trivially_copy_constructible is_nothrow_copy_constructible	Check whether a type has a copy constructor.

Name	Description

`is_move_constructible`
 	
`is_trivially_move_constructible`
 	
`is_nothrow_move_constructible`
	Check whether a type has a move constructor.

`is_assignable`
 	
`is_trivially_assignable`
 	
`is_nothrow_assignable`
	Check whether a type has an assignment operator for a specific argument.

`is_copy_assignable`
 	
`is_trivially_copy_assignable`
 	
`is_nothrow_copy_assignable`
	Check whether a type a copy assignment operator.

`is_move_assignable`
 	
`is_trivially_move_assignable`
 	
`is_nothrow_move_assignable`
	Check whether a type has a move assignment operator.

`is_destructible`
 	
`is_trivially_destructible`
 	
`is_nothrow_destructible`
	Check whether a type has a non-deleted destructor.

`has_virtual_destructor`	Checks whether a type has a virtual destructor.

`is_swappable_with`
 	
`is_swappable`
 	
`is_nothrow_swappable_with`
 	
`is_nothrow_swappable`
	Check whether objects of the same type or objects of two different types can be swapped.

Table 5.3

Except for the last sub-set, which was introduced in C++17, the others are available in C++11. Each kind of these type traits has multiple variants, including ones for checking operations that are trivial or declared as non-throwing with the `noexcept` specifier.

Now let's look at type traits that allow us to query for relationships between types.

Querying type relationships

In this category, we can find several type traits that help to query relationships between types. These type traits are as follows:

Name	C++ Version	Description
`is_same`	C++11	Checks whether two types are the same, including possible cv-qualifiers.
`is_base_of`	C++11	Checks whether one type is derived from another type.
`is_convertible`	C++11	Check whether a type can be converted to another type.
`is_nothrow_convertible`	C++14	
`is_invocable`	C++17	Check whether a type can be invoked with the specified argument types.
`is_invocable_r`		
`is_nothrow_invocable`		
`is_nothrow_invocable_r`		
`is_layout_compatible`	C++20	Checks whether two types have compatible layouts. Two classes are layout-compatible if they are the same type (ignoring cv-qualifiers), or their common initial sequence contains all the non-static data members and bit fields or are enumerations with the same underlying type.
`is_pointer_inconvertible_base_of`	C++20	Checks whether a type is a pointer-inconvertible base of another type.

Table 5.4

Of these type traits, perhaps the most used one is `std::is_same`. This type trait is very useful in determining whether two types are the same. Keep in mind that this type trait takes into account the `const` and `volatile` qualifiers; therefore, `int` and `int const`, for instance, are not the same type.

We can use this type trait to extend the implementation of the `as_string` function shown earlier. Remember that if you called it with the arguments `true` or `false` it prints `1` or `0`, and not `true`/`false`. We can add an explicit check for the `bool` type and return a string containing one of these two values, shown as follows:

```cpp
template <typename T>
std::string as_string(T value)
{
    if constexpr (std::is_null_pointer_v<T>)
        return "null";
    else if constexpr (std::is_same_v<T, bool>)
        return value ? "true" : "false";
    else if constexpr (std::is_arithmetic_v<T>)
        return std::to_string(value);
    else
        static_assert(always_false<T>);
}


std::cout << as_string(true) << '\n';     // prints true
std::cout << as_string(false) << '\n';    // prints false
```

All the type traits seen so far are used to query some kind of information about types. In the next sections, we will see type traits that perform modifications on types.

Modifying cv-specifiers, references, pointers, or a sign

The type traits that are performing transformations on types are also called metafunctions. These type traits provided a member type (typedef) called type that represents the transformed type. This category of type traits includes the following:

Name	Description
add_cv add_const add_volatile	Add the const, volatile, or both specifiers to a type.
remove_cv remove_const remove_volatile	Remove the const, volatile, or both specifiers from a type.
add_lvalue_reference add_rvalue_reference	Add an lvalue or rvalue reference to a type.
remove_reference	Removes a reference (either lvalue or rvalue) from a type.

Name	Description
remove_cvref	Removes the const and volatile specifiers as well as lvalue or rvalue references from a type. It combines the remove_cv and remove_reference traits.
add_pointer	Adds a pointer to a type.
remove_pointer	Removes a pointer from a type.
make_signed make_unsigned	Make an integral type (except for bool) or an enumeration type either signed or unsigned. The supported integral types are short, int, long, long long, char, wchar_t, char8_t, char16_t, and char32_t.
remove_extent remove_all_extents	Remove one or all extents from an array type.

Table 5.5

With the exception of remove_cvref, which was added in C++20, all the other type traits listed in this table are available in C++11. These are not all the metafunctions from the standard library. More are listed in the next section.

Miscellaneous transformations

Apart from the metafunctions previously listed, there are other type traits performing type transformations. The most important of these are listed in the following table:

Name	C++ version	Description
enable_if	C++11	Enables the removal of a function overload or template specialization from overload resolution.
conditional	C++11	Defines the member type called type to be one of two possible types by selecting them based on a compile-time Boolean condition.
decay	C++11	Applies transformations on a type (array-to-pointer for array types, lvalue-to-rvalue for reference types, and function-to-pointer to function types), removes const and volatile qualifiers, and uses the resulting member typedef type as its own member typedef type.
common_type	C++11	Determines the common type from a group of types.

Name	C++ version	Description
`common_reference`	C++20	Determines the common reference type from a group of types.
`underlaying_type`	C++11	Determines the underlying type of an enumeration type.
`void_t`	C++17	A type alias mapping a sequence of types to the `void` type.
`type_identity`	C++20	Provides the member typedef `type` as an alias for the type argument `T`.

Table 5.6

From this list, we have already discussed `enable_if`. There are some other type traits here that are worth exemplifying. Let's first look at `std::decay` and for this purpose, let's consider the following slightly changed implementation of the `as_string` function:

```
template <typename T>
std::string as_string(T&& value)
{
    if constexpr (std::is_null_pointer_v<T>)
        return "null";
    else if constexpr (std::is_same_v<T, bool>)
        return value ? "true" : "false";
    else if constexpr (std::is_arithmetic_v<T>)
        return std::to_string(value);
    else
        static_assert(always_false<T>);
}
```

The only change is the way we pass arguments to the function. Instead of passing by value, we pass by rvalue reference. If you remember from *Chapter 4, Advanced Template Concepts*, this is a forwarding reference. We can still make calls passing rvalues (such as literals) but passing lvalues will trigger compiler errors:

```
std::cout << as_string(true) << '\n';   // OK
std::cout << as_string(42) << '\n';     // OK

bool f = true;
```

```
std::cout << as_string(f) << '\n';        // error

int n = 42;
std::cout << as_string(n) << '\n';        // error
```

The last two calls are triggering the `static_assert` statement to fail. The actual type template arguments are `bool&` and `int&`. Therefore `std::is_same<bool, bool&>` will initialize the `value` member with `false`. Similarly, `std::is_arithmetic<int&>` will do the same. In order to evaluate these types, we need to ignore references and the `const` and `volatile` qualifiers. The type trait that helps us do so is `std::decay`, which performs several transformations, as described in the previous table. Its conceptual implementation is the following:

```
template <typename T>
struct decay
{
private:
    using U = typename std::remove_reference_t<T>;
public:
    using type = typename std::conditional_t<
        std::is_array_v<U>,
        typename std::remove_extent_t<U>*,
        typename std::conditional_t<
            std::is_function<U>::value,
            typename std::add_pointer_t<U>,
            typename std::remove_cv_t<U>
        >
    >;
};
```

From this snippet, we can see that `std::decay` is implemented with the help of other metafunctions, including `std::conditional`, which is key for selecting between one type or another based on a compile-time expression. Actually, this type trait is used multiple times, which is something you can do if you need to make a selection based on multiple conditions.

With the help of std::decay, we can modify the implementation of the as_string function, stripping reference, and cv-qualifiers:

```
template <typename T>
std::string as_string(T&& value)
{
    using value_type = std::decay_t<T>;

    if constexpr (std::is_null_pointer_v<value_type>)
        return "null";
    else if constexpr (std::is_same_v<value_type, bool>)
        return value ? "true" : "false";
    else if constexpr (std::is_arithmetic_v<value_type>)
        return std::to_string(value);
    else
        static_assert(always_false<T>);
}
```

By changing the implementation as shown here, we made the previous calls to as_string that failed to compile without any more errors.

In the implementation of std::decay we saw the repetitive use of std::conditional. This is a metafunction that is fairly easy to use and can help to simplify many implementations. In *Chapter 2, Template Fundamentals*, in the section *Defining alias templates*, we saw an example where we built a list type called list_t. This had a member alias template called type that was aliasing either the template type T, if the size of the list was 1, or std::vector<T>, if it was higher. Let's look at the snippet again:

```
template <typename T, size_t S>
struct list
{
    using type = std::vector<T>;
};

template <typename T>
struct list<T, 1>
{
    using type = T;
```

```
};

template <typename T, size_t S>
using list_t = typename list<T, S>::type;
```

This implementation can be greatly simplified with the help of `std::conditional` as follows:

```
template <typename T, size_t S>
using list_t =
    typename std::conditional<S ==
                  1, T, std::vector<T>>::type;
```

There is no need to rely on class template specialization to define such a list type. The entire solution can be reduced to defining an alias template. We can verify this works as expected with some `static_assert` statements, as follows:

```
static_assert(std::is_same_v<list_t<int, 1>, int>);
static_assert(std::is_same_v<list_t<int, 2>,
                             std::vector<int>>);
```

Exemplifying the use of each of the standard type traits is beyond the scope of this book. However, the next section of this chapter provides more complex examples that require the use of several standard type traits.

Seeing real-world examples of using type traits

In the previous section of the chapter, we have explored the various type traits that the standard library provides. It is difficult and unnecessary to find examples for each and every type trait. However, it is worth showcasing some examples where multiple type traits can be used for solving a problem. We will do this next.

Implementing a copy algorithm

The first example problem we will take a look at is a possible implementation for the
`std::copy` standard algorithm (from the `<algorithm>` header). Keep in mind that
what we will see next is not the actual implementation but a possible one that helps us
learn more about the use of type traits. The signature of this algorithm is as follows:

```
template <typename InputIt, typename OutputIt>
constexpr OutputIt copy(InputIt first, InputIt last,
                        OutputIt d_first);
```

As a note, this function is `constexpr` only in C++20, but we can discuss it in this
context. What it does is copy all the elements in the range [`first`, `last`) to another
range that begins with `d_first`. There is also an overload that takes an execution policy,
and a version, `std::copy_if`, that copies all the elements that match a predicate,
but these are not important for our example. A straightforward implementation of this
function is the following:

```
template <typename InputIt, typename OutputIt>
constexpr OutputIt copy(InputIt first, InputIt last,
                        OutputIt d_first)
{
    while (first != last)
    {
        *d_first++ = *first++;
    }
    return d_first;
}
```

However, there are cases when this implementation can be optimized by simply copying
memory. However, there are some conditions that must be met for this purpose:

- Both iterator types, `InputIt` and `OutputIt`, must be pointers.
- Both template parameters, `InputIt` and `OutputIt`, must point to the same type
  (ignoring cv-qualifiers).
- The type pointed by `InputIt` must have a trivial copy assignment operator.

We can check these conditions with the following standard type traits:

- `std::is_same` (and the `std::is_same_v` variable) to check that two types are the same.

- `std::is_pointer` (and the `std::is_pointer_v` variable) to check that a type is a pointer type.

- `std::is_trivially_copy_assignable` (and the `std::is_trivially_copy_assignable_v` variable) to check whether a type has a trivial copy assignment operator.

- `std::remove_cv` (and the `std::remove_cv_t` alias template) to remove cv-qualifiers from a type.

Let's see how we can implement this. First, we need to have a primary template with the generic implementation, and then a specialization for pointer types with the optimized implementation. We can do this using class templates with member function templates as shown next:

```cpp
namespace detail
{
    template <bool b>
    struct copy_fn
    {
        template<typename InputIt, typename OutputIt>
        constexpr static OutputIt copy(InputIt first,
                                       InputIt last,
                                       OutputIt d_first)
        {
            while (first != last)
            {
                *d_first++ = *first++;
            }
            return d_first;
        }
    };

    template <>
    struct copy_fn<true>
    {
```

```
    template<typename InputIt, typename OutputIt>
    constexpr static OutputIt* copy(
        InputIt* first, InputIt* last,
        OutputIt* d_first)
    {
        std::memmove(d_first, first,
                     (last - first) * sizeof(InputIt));
        return d_first + (last - first);
    }
  };
}
```

To copy memory between a source and a destination we use `std::memmove` here, which copies data even if objects overlap. These implementations are provided in a namespace called `detail`, because they are implementation details that are used in turn by the `copy` function and not directly by the user. The implementation of this generic `copy` algorithm could be as follows:

```
template<typename InputIt, typename OutputIt>
constexpr OutputIt copy(InputIt first, InputIt last,
                        OutputIt d_first)
{
    using input_type = std::remove_cv_t<
        typename std::iterator_traits<InputIt>::value_type>;
    using output_type = std::remove_cv_t<
        typename std::iterator_traits<OutputIt>::value_type>;

    constexpr bool opt =
        std::is_same_v<input_type, output_type> &&
        std::is_pointer_v<InputIt> &&
        std::is_pointer_v<OutputIt> &&
        std::is_trivially_copy_assignable_v<input_type>;

    return detail::copy_fn<opt>::copy(first, last, d_first);
}
```

You can see here that the decision to select one specialization or the other is based on a `constexpr` Boolean value that is determined using the aforementioned type traits. Examples of using this `copy` function are shown in the next snippet:

```cpp
std::vector<int> v1{ 1, 2, 3, 4, 5 };
std::vector<int> v2(5);

// calls the generic implementation
copy(std::begin(v1), std::end(v1), std::begin(v2));

int a1[5] = { 1,2,3,4,5 };
int a2[5];

// calls the optimized implementation
copy(a1, a1 + 5, a2);
```

Keep in mind that this is not the real definition of the generic algorithm `copy` you will find in standard library implementations, which are further optimized. However, this was a good example to demonstrate how to use type traits for a real-world problem.

For simplicity, I have defined the `copy` function in what appears to be the global namespace. This is a bad practice. In general, code, especially in libraries, is grouped in namespaces. In the source code on GitHub that accompanies the book, you will find this function defined in a namespace called `n520` (this is just a unique name, nothing relevant to the topic). When calling the `copy` function that we have defined, we would actually need to use the fully qualified name (that includes the namespace name) such as the following:

```cpp
n520::copy(std::begin(v1), std::end(v1), std::begin(v2));
```

Without this qualification, a process called **Argument-Dependent Lookup (ADL)** would kick in. This would result in resolving the call to `copy` to the `std::copy` function because the arguments we pass are found in the `std` namespace. You can read more about ADL at `https://en.cppreference.com/w/cpp/language/adl`.

Now, let's look at another example.

Building a homogenous variadic function template

For the second example, we want to build a variadic function template that can only take arguments of the same type or types that can be implicitly converted to a common one. Let's start with the following skeleton definition:

```
template<typename... Ts>
void process(Ts&&... ts) {}
```

The problem with this is that all of the following function calls work (keep in mind that the body of this function is empty so there will be no errors due to performing operations unavailable on some types):

```
process(1, 2, 3);
process(1, 2.0, '3');
process(1, 2.0, "3");
```

In the first example, we pass three `int` values. In the second example, we pass an `int`, a `double`, and a `char`; both `int` and `char` are implicitly convertible to `double`, so this should be all right. However, in the third example, we pass an `int`, a `double`, and a `char const*`, and this last type is not implicitly convertible to either `int` or `double`. Therefore, this last call is supposed to trigger a compiler error but does not.

In order to do so, we need to ensure that when a common type for the function arguments is not available, the compiler will generate an error. To do so, we can use a `static_assert` statement or `std::enable_if` and SFINAE. However, we do need to figure out whether a common type exists or not. This is possible with the help of the `std::common_type` type trait.

The `std::common_type` is a metafunction that defines the common type among all of its type arguments that all the types can be implicitly converted to. Therefore `std::common_type<int, double, char>::type` will alias the `double` type. Using this type trait, we can build another type trait that tells us whether a common type exists. A possible implementation is as follows:

```
template <typename, typename... Ts>
struct has_common_type : std::false_type {};

template <typename... Ts>
struct has_common_type<
          std::void_t<std::common_type_t<Ts...>>,
          Ts...>
```

```
    : std::true_type {};

template <typename... Ts>
constexpr bool has_common_type_v =
    sizeof...(Ts) < 2 ||
    has_common_type<void, Ts...>::value;
```

You can see in this snippet that we base the implementation on several other type traits. First, there is the `std::false_type` and `std::true_type` pair. These are type aliases for `std::bool_constant<false>` and `std::bool_constant<true>` respectively. The `std::bool_constant` class is available in C++17 and is, in turn, an alias template for a specialization of the `std::integral_constant` class for the `bool` type. This last class template wraps a static constant of the specified type. Its conceptual implementation looks as follows (although some operations are also provided):

```
template<class T, T v>
struct integral_constant
{
    static constexpr T value = v;
    using value_type = T;
};
```

This helps us simplify the definition of type traits that need to define a Boolean compile-time value, as we saw in several cases in this chapter.

A third type trait used in the implementation of the `has_common_type` class is `std::void_t`. This type trait defines a mapping between a variable number of types and the `void` type. We use this to build a mapping between the common type, if one exists, and the `void` type. This enables us to leverage SFINAE for the specialization of the has_common_type class template.

Finally, a variable template called `has_common_type_v` is defined to ease the use of the `has_common_type` trait.

All these can be used to modify the definition of the `process` function template to ensure it only allows arguments of a common type. A possible implementation is shown next:

```
template<typename... Ts,
         typename = std::enable_if_t<
                        has_common_type_v<Ts...>>>
void process(Ts&&... ts)
{ }
```

As a result of this, calls such as `process(1, 2.0, "3")` will produce a compiler error because there is no overloaded `process` function for this set of arguments.

As previously mentioned, there are different ways to use the `has_common_type` trait to achieve the defined goal. One of these, using `std::enable_if`, was shown here, but we can also use `static_assert`. However, a much better approach can be taken with the use of concepts, which we will see in the next chapter.

Summary

This chapter explored the concept of type traits, which are small classes that define meta-information about types or transformation operations for types. We started by looking at how type traits can be implemented and how they help us. Next, we learned about **SFINAE**, which stands for **Substitution Failure Is Not An Error**. This is a technique that enables us to provide constraints for template parameters.

We then saw how this purpose can be achieved better with `enable_if` and `constexpr if`, in C++17. In the second part of the chapter, we looked at the type traits available in the standard library and demonstrated how to use some of them. We ended the chapter with a couple of real-world examples where we used multiple type traits to solve a particular problem.

In the next chapter, we continue the topic of constraining the template parameters by learning about the C++20 concepts and constraints.

Questions

1. What are type traits?
2. What is SFINAE?
3. What is `constexpr if`?
4. What does `std::is_same` do?
5. What does `std::conditional` do?

Further reading

- *C++ Type traits*, John Maddock and Steve Cleary, `https://cs.brown.edu/~jwicks/boost/libs/type_traits/cxx_type_traits.htm`
- *N4861 Post-Prague 2020 C++ working draft*, `https://github.com/cplusplus/draft/releases/tag/n4861`
- *What is ADL?*, Arthur O'Dwyer, `https://quuxplusone.github.io/blog/2019/04/26/what-is-adl/`

6
Concepts and Constraints

The C++20 standard provides a series of significant improvements to template metaprogramming with concepts and constraints. A **constraint** is a modern way to define requirements on template parameters. A **concept** is a set of named constraints. Concepts provide several benefits to the traditional way of writing templates, mainly improved readability of code, better diagnostics, and reduced compilation times.

In this chapter, we will address the following topics:

- Understanding the need for concepts
- Defining concepts
- Exploring requires expressions
- Composing constraints
- Learning about the ordering of templates with constraints
- Constraining non-template member functions
- Constraining class templates
- Constraining variable templates and template aliases
- Learning more ways to specify constraints

- Using concepts to constrain auto parameters
- Exploring the standard concepts library

By the end of this chapter, you will have a good understanding of the C++20 concepts, and an overview of what concepts the standard library provides.

We will start the chapter by discussing what led to the development of concepts and what their main benefits are.

Understanding the need for concepts

As briefly mentioned in the introduction to this chapter, there are some important benefits that concepts provide. Arguably, the most important ones are code readability and better error messages. Before we look at how to use concepts, let's revisit an example we saw previously and see how it stands in relation to these two programming aspects:

```
template <typename T>
T add(T const a, T const b)
{
    return a + b;
}
```

This simple function template takes two arguments and returns their sum. In fact, it does not return the sum, but the result of applying the plus operator to the two arguments. A user-defined type can overload this operator and perform some particular operation. The term *sum* only makes sense when we discuss mathematical types, such as integral types, floating-point types, the `std::complex` type, matrix types, vector types, etc.

For a string type, for instance, the plus operator can mean concatenation. And for most types, its overloading does not make sense at all. Therefore, just by looking at the declaration of the function, without inspecting its body, we cannot really say what this function may accept as input and what it does. We can call this function as follows:

```
add(42, 1);         // [1]
add(42.0, 1.0);     // [2]
add("42"s, "1"s);   // [3]
add("42", "1");     // [4] error: cannot add two pointers
```

The first three calls are all good; the first call adds two integers, the second adds two `double` values, and the third concatenates two `std::string` objects. However, the fourth call will produce a compiler error because `const char *` is substituted for the `T` type template parameter, and the plus operator is not overloaded for pointer types.

The intention for this `add` function template is to allow passing only values of arithmetic types, that is, integer and floating-point types. Before C++20, we could do this in several ways.

One way is to use `std::enable_if` and SFINAE, as we saw in the previous chapter. Here is such an implementation:

```
template <typename T,
    typename = typename std::enable_if_t
        <std::is_arithmetic_v<T>>>
T add(T const a, T const b)
{
    return a + b;
}
```

The first thing to notice here is that the readability has decreased. The second type template parameter is difficult to read and requires good knowledge of templates to understand. However, this time, both the calls on the lines marked with [3] and [4] are producing a compiler error. Different compilers are issuing different error messages. Here are the ones for the three major compilers:

- In **VC++ 17**, the output is:

  ```
  error C2672: 'add': no matching overloaded function found
  error C2783: 'T add(const T,const T)': could not deduce
  template argument for '<unnamed-symbol>'
  ```

- In **GCC 12**, the output is:

  ```
  prog.cc: In function 'int main()':
  prog.cc:15:8: error: no matching function for call
  to 'add(std::__cxx11::basic_string<char>, std::__
  cxx11::basic_string<char>)'
     15 |      add("42"s, "1"s);
        |      ~~~^~~~~~~~~~~~~~
  prog.cc:6:6: note: candidate: 'template<class T, class> T
  add(T, T)'
  ```

```
  6 |      T add(T const a, T const b)
    |         ^~~
```

prog.cc:6:6: note: template argument deduction/
substitution failed:

In file included from /opt/wandbox/gcc-head/include/
c++/12.0.0/bits/move.h:57,

                from /opt/wandbox/gcc-head/include/
c++/12.0.0/bits/nested_exception.h:40,

                from /opt/wandbox/gcc-head/include/
c++/12.0.0/exception:154,

                from /opt/wandbox/gcc-head/include/
c++/12.0.0/ios:39,

                from /opt/wandbox/gcc-head/include/
c++/12.0.0/ostream:38,

                from /opt/wandbox/gcc-head/include/
c++/12.0.0/iostream:39,

                from prog.cc:1:
/opt/wandbox/gcc-head/include/c++/12.0.0/type_traits: In
substitution of 'template<bool _Cond, class _Tp> using
enable_if_t = typename std::enable_if::type [with bool _
Cond = false; _Tp = void]':

prog.cc:5:14:   required from here

/opt/wandbox/gcc-head/include/c++/12.0.0/type_
traits:2603:11: error: no type named 'type' in 'struct
std::enable_if<false, void>'

```
 2603 |      using enable_if_t = typename enable_if<_Cond,
_Tp>::type;
      |                        ^~~~~~~~~~~
```

- In **Clang 13**, the output is:

```
prog.cc:15:5: error: no matching function for call to
'add'
    add("42"s, "1"s);
    ^~~
```

prog.cc:6:6: note: candidate template ignored:
requirement 'std::is_arithmetic_v<std::string>' was not
satisfied [with T = std::string]

```
  T add(T const a, T const b)
    ^
```

The error message in GCC is very verbose, and VC++ doesn't say what the reason for failing to match the template argument is. Clang does, arguably, a better job at providing an understandable error message.

Another way to define restrictions for this function, prior to C++20, is with the help of a static_assert statement, as shown in the following snippet:

```
template <typename T>
T add(T const a, T const b)
{
    static_assert(std::is_arithmetic_v<T>,
                  "Arithmetic type required");
    return a + b;
}
```

With this implementation, however, we returned to the original problem that just by looking at the declaration of the function, we wouldn't know what kind of parameters it would accept, provided that any restriction exists. The error messages, on the other hand, are as follows:

- In **VC++ 17**:

```
error C2338: Arithmetic type required
main.cpp(157): message : see reference to function
template instantiation 'T add<std::string>(const T,const
T)' being compiled
    with
    [
        T=std::string
    ]
```

- In **GCC 12**:

```
prog.cc: In instantiation of 'T add(T, T) [with T =
std::__cxx11::basic_string<char>]':
prog.cc:15:8:    required from here
prog.cc:7:24: error: static assertion failed: Arithmetic
type required
    7 |       static_assert(std::is_arithmetic_v<T>,
"Arithmetic type required");
```

```
        |                           ^
                         ~~~~~^~~~~~~~~~~~~~~~~~~
prog.cc:7:24: note: 'std::is_arithmetic_v<std::__
cxx11::basic_string<char> >' evaluates to false
```

- In **Clang 13**:

```
prog.cc:7:5: error: static_assert failed due to
requirement 'std::is_arithmetic_v<std::string>'
"Arithmetic type required"
    static_assert(std::is_arithmetic_v<T>, "Arithmetic
type required");
    ^
                  ~~~~~~~~~~~~~~~~~~~~~~~~
prog.cc:15:5: note: in instantiation of function template
specialization 'add<std::string>' requested here
    add("42"s, "1"s);
    ^
```

The use of the `static_assert` statement results in similar error messages received regardless of the compiler.

We can improve these two discussed aspects (readability and error messages) in C++20 by using constraints. These are introduced with the new `requires` keyword as follows:

```
template <typename T>
requires std::is_arithmetic_v<T>
T add(T const a, T const b)
{
    return a + b;
}
```

The `requires` keyword introduces a clause, called the **requires clause**, that defines the constraints on the template parameters. There are, actually, two alternative syntaxes: one when the requires clause follows the template parameter list, as seen previously, and one when the requires clause follows the function declaration, as shown in the next snippet:

```
template <typename T>
T add(T const a, T const b)
requires std::is_arithmetic_v<T>
{
    return a + b;
}
```

Choosing between these two syntaxes is a matter of personal preference. However, in both cases, the readability is much better than in the pre-C++20 implementations. You know just by reading the declaration that the T type template parameter must be of an arithmetic type. Also, this implies that the function is simply adding two numbers. You don't really need to see the definition to know that. Let's see how the error message changes when we call the function with invalid arguments:

- In **VC++ 17**:

```
error C2672: 'add': no matching overloaded function found
error C7602: 'add': the associated constraints are not
satisfied
```

- In **GCC 12**:

```
prog.cc: In function 'int main()':
prog.cc:15:8: error: no matching function for call
to 'add(std::__cxx11::basic_string<char>, std::__
cxx11::basic_string<char>)'
   15 |      add("42"s, "1"s);
      |          ~~~^~~~~~~~~~~~~
prog.cc:6:6: note: candidate: 'template<class
T> requires  is_arithmetic_v<T> T add(T, T)'
    6 |      T add(T const a, T const b)
      |        ^~~
prog.cc:6:6: note:    template argument deduction/
substitution failed:
prog.cc:6:6: note: constraints not satisfied
prog.cc: In substitution of 'template<class
T> requires  is_arithmetic_v<T> T add(T, T) [with T =
std::__cxx11::basic_string<char>]':
prog.cc:15:8:    required from here
prog.cc:6:6:    required by the constraints of
'template<class T> requires  is_arithmetic_v<T> T add(T,
T)'
prog.cc:5:15: note: the expression 'is_arithmetic_v<T>
[with T = std::__cxx11::basic_string<char, std::char_
traits<char>, std::allocator<char> >]' evaluated to
'false'
    5 | requires std::is_arithmetic_v<T>
      |          ~~~~~^~~~~~~~~~~~~~~~~~~~~
```

- In **Clang 13**:

```
prog.cc:15:5: error: no matching function for call to
'add'
    add("42"s, "1"s);
    ^~~

prog.cc:6:6: note: candidate template ignored:
constraints not satisfied [with T = std::string]
   T add(T const a, T const b)
     ^

prog.cc:5:10: note: because 'std::is_arithmetic_
v<std::string>' evaluated to false
requires std::is_arithmetic_v<T>
         ^
```

The error messages follow the same patterns seen already: GCC is too verbose, VC++ is missing essential information (the constraint that is not met), while Clang is more concise and better pinpoints the cause of the error. Overall, there is an improvement in the diagnostic messages, although there is still room for improvement.

A constraint is a predicate that evaluates to true or false at compile-time. The expression used in the previous example, `std::is_arithmetic_v<T>`, is simply using a standard type trait (which we saw in the previous chapter). However, these are different kinds of expressions that can be used in a constraint, and we will learn about them later in this chapter.

In the next section, we look at how to define and use named constraints.

Defining concepts

The constraints seen previously are nameless predicates defined in the places they are used. Many constraints are generic and can be used in multiple places. Let's consider the following example of a function similar to the add function. This function performs the multiplication of arithmetic values and is shown next:

```
template <typename T>
requires std::is_arithmetic_v<T>
T mul(T const a, T const b)
{
    return a * b;
}
```

The same requires clause seen with the add function is present here. To avoid this repetitive code, we can define a name constraint that can be reused in multiple places. A named constraint is called a **concept**. A concept is defined with the new `concept` keyword and template syntax. Here is an example:

```
template<typename T>
concept arithmetic = std::is_arithmetic_v<T>;
```

Even though they are assigned a Boolean value, concept names should not contain verbs. They represent requirements and are used as attributes or qualifiers on template parameters. Therefore, you should prefer names such as *arithmetic*, *copyable*, *serializable*, *container*, and more, and not *is_arithmetic*, *is_copyable*, *is_serializable*, and *is_container*. The previously defined arithmetic concept can be used as follows:

```
template <arithmetic T>
T add(T const a, T const b) { return a + b; }


template <arithmetic T>
T mul(T const a, T const b) { return a * b; }
```

You can see from this snippet that the concept is used instead of the `typename` keyword. It qualifies the `T` type with the arithmetic quality, meaning that only the types that satisfy this requirement can be used as template arguments. The same arithmetic concept can be defined with a different syntax, shown in the following snippet:

```
template<typename T>
concept arithmetic = requires { std::is_arithmetic_v<T>; };
```

This uses a *requires expression*. A requires expression uses curly branches, { }, whereas a *requires clause* does not. A requires expression can contain a sequence of requirements of different kinds: simple requirements, type requirements, compound requirements, and nested requirements. The one seen here is a simple requirement. For the purpose of defining this particular concept, this syntax is more complicated but has the same final effect. However, in some cases, complex requirements are needed. Let's look at an example.

Consider the case when we want to define a template that should only take container types for an argument. Before concepts were available, this could have been solved with the help of a type trait and SFINAE or a `static_assert` statement, as we saw at the beginning of this chapter. However, a container type is not really easy to define formally. We can do it based on some properties of the standard containers:

- They have the member types `value_type`, `size_type`, `allocator_type`, `iterator`, and `const_iterator`.

- They have the member function `size` that returns the number of elements in the container.

- They have the member functions `begin`/`end` and `cbegin`/`cend` that return iterators and constant iterators to the first and one-past-the-last element in the container.

With the knowledge accumulated from *Chapter 5*, *Type Traits and Conditional Compilation*, we can define an `is_containter` type trait as follows:

```cpp
template <typename T, typename U = void>
struct is_container : std::false_type {};

template <typename T>
struct is_container<T,
    std::void_t<typename T::value_type,
                typename T::size_type,
                typename T::allocator_type,
                typename T::iterator,
                typename T::const_iterator,
                decltype(std::declval<T>().size()),
                decltype(std::declval<T>().begin()),
                decltype(std::declval<T>().end()),
                decltype(std::declval<T>().cbegin()),
                decltype(std::declval<T>().cend())>>
    : std::true_type{};

template <typename T, typename U = void>
constexpr bool is_container_v = is_container<T, U>::value;
```

We can verify with the help of `static_assert` statements that the type trait correctly identifies container types. Here is an example:

```
struct foo {};

static_assert(!is_container_v<foo>);
static_assert(is_container_v<std::vector<foo>>);
```

Concepts make writing such a template constraint much easier. We can employ the concept syntax and requires expressions to define the following:

```
template <typename T>
concept container = requires(T t)
{
    typename T::value_type;
    typename T::size_type;
    typename T::allocator_type;
    typename T::iterator;
    typename T::const_iterator;
    t.size();
    t.begin();
    t.end();
    t.cbegin();
    t.cend();
};
```

This definition is both shorter and more readable. It uses both simple requirements, such as `t.size()`, and type requirements, such as `typename T::value_type`. It can be used to constrain template parameters in the manner seen previously but can also be used with the `static_assert` statements (since constraints evaluate to a compile-time Boolean value):

```
struct foo{};

static_assert(!container<foo>);
static_assert(container<std::vector<foo>>);

template <container C>
void process(C&& c) {}
```

In the following section, we will explore in depth the various kinds of requirements that can be used in requires expressions.

Exploring requires expressions

A requires expression may be a complex expression, as seen earlier in the example with the container concept. The actual form of a requires expression is very similar to function syntax and is as follows:

```
requires (parameter-list) { requirement-seq }
```

The `parameter-list` is a comma-separated list of parameters. The only difference from a function declaration is that default values are not allowed. However, the parameters that are specified in this list do not have storage, linkage, or lifetime. The compiler does not allocate any memory for them; they are only used to define requirements. However, they do have a scope, and that is the closing curly brace of the requires expression.

The `requirements-seq` is a sequence of requirements. Each such requirement must end with a semicolon, like any statement in C++. There are four types of requirements:

- Simple requirements
- Type requirements
- Compound requirements
- Nested requirements

These requirements may refer to the following:

- Template parameters that are in scope
- Local parameters introduced in the parameter list of the requires expression
- Any other declaration that is visible from the enclosing context

In the following subsections, we will explore all the mentioned types of requirements. In the beginning, we'll look at the simple requirements.

Simple requirements

A **simple requirement** is an expression that is not evaluated but only checked for correctness. The expression must be valid for the requirement to be evaluated to the value `true`. The expression must not start with the `requires` keyword as that defines a nested requirement (which will be discussed later).

We already saw examples of simple statements when we defined the `arithmetic` and `container` concepts earlier. Let's see a few more:

```
template<typename T>
concept arithmetic = requires
{
    std::is_arithmetic_v<T>;
};

template <typename T>
concept addable = requires(T a, T b)
{
    a + b;
};

template <typename T>
concept logger = requires(T t)
{
    t.error("just");
    t.warning("a");
    t.info("demo");
};
```

The first concept, `arithmetic`, is the same one we defined earlier. The `std::is_arithmetic_v<T>` expression is a simple requirement. Notice that when the parameter list is empty it can be completely omitted, as seen in this case, where we only check that the T type template parameter is an arithmetic type.

The `addable` and `logger` concepts both have a parameter list because we are checking operations on values of the T type. The expression a + b is a simple requirement, as the compiler just checks that the plus operator is overloaded for the T type. In the last example, we make sure that the T type has three member functions called `error`, `warning`, and `info` that take a single parameter of the `const char*` type or some type that can be constructed from `const char*`. Keep in mind that the actual values passed as arguments have no importance since these calls are never performed; they are only checked for correctness.

Let's elaborate briefly on the last example and consider the following snippet:

```
template <logger T>
void log_error(T& logger)
{}


struct console_logger
{
    void error(std::string_view text){}
    void warning(std::string_view text) {}
    void info(std::string_view text) {}
};


struct stream_logger
{
    void error(std::string_view text, bool = false) {}
    void warning(std::string_view text, bool = false) {}
    void info(std::string_view text, bool) {}
};
```

The `log_error` function template requires an argument of a type that meets the `logger` requirements. We have two classes, called `console_logger` and `stream_logger`. The first meets the `logger` requirements, but the second does not. That is because the `info` function cannot be invoked with a single argument of type `const char*`. This function also requires a second, Boolean, argument. The first two methods, `error` and `warning`, define a default value for the second argument, so they can be invoked with calls such as `t.error("just")` and `warning("a")`.

However, because of the third member function, `stream_logger` is not a log class that meets the expected requirements and, therefore, cannot be used with the `log_error` function. The use of `console_logger` and `stream_logger` is exemplified in the following snippet:

```
console_logger cl;
log_error(cl);      // OK


stream_logger sl;
log_error(sl);      // error
```

In the next section, we look at the second category of requirements, type requirements.

Type requirements

Type requirements are introduced with the keyword `typename` followed by the name of a type. We have already seen several examples when we defined the `container` constraint. The name of the type must be valid for the requirement to be true. Type requirements can be used for several purposes:

- To verify that a nested type exists (such as in `typename T::value_type;`)
- To verify that a class template specialization names a type
- To verify that an alias template specialization names a type

Let's see several examples to learn how to use type requirements. In the first example, we check whether a type contains the inner types, `key_type` and `value_type`:

```cpp
template <typename T>
concept KVP = requires
{
    typename T::key_type;
    typename T::value_type;
};


template <typename T, typename V>
struct key_value_pair
{
    using key_type = T;
    using value_type = V;

    key_type     key;
    value_type   value;
};


static_assert(KVP<key_value_pair<int, std::string>>);
static_assert(!KVP<std::pair<int, std::string>>);
```

The type, `key_value_pair<int, std::string>`, satisfies these type requirements, but `std::pair<int, std::string>` does not. The `std::pair` type does have inner types, but they are called `first_type` and `second_type`.

In the second example, we check whether a class template specialization names a type. The class template is `container`, and the specialization is `container<T>`:

```
template <typename T>
requires std::is_arithmetic_v<T>
struct container
{ /* ... */ };


template <typename T>
concept containerizeable = requires {
    typename container<T>;
};


static_assert(containerizeable<int>);
static_assert(!containerizeable<std::string>);
```

In this snippet, `container` is a class template that can only be specialized for arithmetic types, such as `int`, `long`, `float`, or `double`. Therefore, specializations such as `container<int>` exist, but `container<std::string>` does not. The `containerizeable` concept specifies a requirement for a type T to define a valid specialization of `container`. Therefore, `containerizeable<int>` is true, but `containerizeable<std::string>` is false.

Now that we have understood simple requirements and type requirements it is time to explore the more complex category of requirements. The first to look at is compound requirements.

Compound requirements

Simple requirements allow us to verify that an expression is valid. However, sometimes we need to verify some properties of an expression not just that it is valid. This can include whether an expression does not throw exceptions or requirements on the result type (such as the return type of a function). The general form is the following:

```
{ expression } noexcept -> type_constraint;
```

Both the noexcept specification and the type_constraint (with the leading ->) are optional. The substitution process and the checking of the constraints occur as follows:

1. The template arguments are substituted in the expression.

2. If noexcept is specified, then the expression must not throw exceptions; otherwise, the requirement is false.

3. If the type constraint is present, then the template arguments are also substituted into type_contraint and decltype((expression)) must satisfy the conditions imposed by type_constraint; otherwise, the requirement is false.

We will discuss a couple of examples to learn how to use compound requirements. In the first example, we check whether a function is marked with the noexcept specifier:

```cpp
template <typename T>
void f(T) noexcept {}

template <typename T>
void g(T) {}

template <typename F, typename ... T>
concept NonThrowing = requires(F && func, T ... t)
{
    {func(t...)} noexcept;
};

template <typename F, typename ... T>
    requires NonThrowing<F, T...>
void invoke(F&& func, T... t)
{
    func(t...);
}
```

In this snippet, there are two function templates: f is declared noexcept; therefore, it shall not throw any exception, and g, which potentially throws exceptions. The NonThrowing concept imposes the requirement that the variadic function of type F must not throw exceptions. Therefore, of the following two invocations, only the first is valid and the second will produce a compiler error:

```
invoke(f<int>, 42);
invoke(g<int>, 42); // error
```

The error messages generated by Clang are shown in the following listing:

```
prog.cc:28:7: error: no matching function for call to 'invoke'
        invoke(g<int>, 42);
        ^~~~~~
prog.cc:18:9: note: candidate template ignored: constraints not
satisfied [with F = void (&)(int), T = <int>]
    void invoke(F&& func, T... t)
         ^
prog.cc:17:16: note: because 'NonThrowing<void (&)(int), int>'
evaluated to false
        requires NonThrowing<F, T...>
                 ^
prog.cc:13:20: note: because 'func(t)' may throw an exception
        {func(t...)} noexcept;
                ^
```

These error messages tell us that the invoke(g<int>, 42) call is not valid because g<int> may throw an exception, which results in NonThrowing<F, T...> to evaluating as false.

For the second example, we will define a concept that provides requirements for timer classes. Specifically, it requires that a function called start exists, that it can be invoked without any parameters, and that it returns void. It also requires that a second function called stop exists, that it can be invoked without any parameters, and that it returns a value that can be converted to long long. The concept is defined as follows:

```
template <typename T>
concept timer = requires(T t)
{
```

```
    {t.start()} -> std::same_as<void>;
    {t.stop()}  -> std::convertible_to<long long>;
};
```

Notice that the type constraint cannot be any compile-time Boolean expression, but an actual type requirement. Therefore, we use other concepts for specifying the return type. Both std::same_as and std::convertible_to are concepts available in the standard library in the <concepts> header. We'll learn more about these in the *Exploring the standard concepts library* section. Now, let's consider the following classes that implement timers:

```
struct timerA
{
    void start() {}
    long long stop() { return 0; }
};


struct timerB
{
    void start() {}
    int stop() { return 0; }
};


struct timerC
{
    void start() {}
    void stop() {}
    long long getTicks() { return 0; }
};


static_assert(timer<timerA>);
static_assert(timer<timerB>);
static_assert(!timer<timerC>);
```

In this example, `timerA` satisfies the timer concept because it contains the two required methods: `start` that returns `void` and `stop` that returns `long long`. Similarly, `timerB` also satisfies the timer concept because it features the same methods, even though `stop` returns an `int`. However, the `int` type is implicitly convertible to the `long long` type; therefore, the type requirement is met. Lastly, `timerC` also has the same methods, but both of them return `void`, which means the type requirement for the return type of `stop` is not met, and therefore, the constraints imposed by the `timer` concept are not satisfied.

The last category of requirements left to look into is nested requirements. We will do this next.

Nested requirements

The last category of requirements is nested requirements. A nested requirement is introduced with the `requires` keyword (remember we mentioned that a simple requirement is a requirement that is not introduced with the `requires` keyword) and has the following form:

```
requires constraint-expression;
```

The expression must be satisfied by the substituted arguments. The substitution of the template arguments into `constraint-expression` is done only to check whether the expression is satisfied or not.

In the following example, we want to define a function that performs addition on a variable number of arguments. However, we want to impose some conditions:

- There is more than one argument.
- All the arguments have the same type.
- The expression `arg1 + arg2 + … + argn` is valid.

To ensure this, we define a concept called `HomogenousRange` as follows:

```
template<typename T, typename... Ts>
inline constexpr bool are_same_v =
    std::conjunction_v<std::is_same<T, Ts>...>;


template <typename ... T>
concept HomogenousRange = requires(T... t)
{
```

```
    (... + t);
    requires are_same_v<T...>;
    requires sizeof...(T) > 1;
};
```

This concept contains one simple requirement and two nested requirements. One nested requirement uses the `are_same_v` variable template whose value is determined by the conjunction of one or more type traits (`std::is_same`), and the other, the compile-time Boolean expression `size...(T) > 1`.

Using this concept, we can define the `add` variadic function template as follows:

```
template <typename ... T>
requires HomogenousRange<T...>
auto add(T&&... t)
{
    return (... + t);
}


add(1, 2);    // OK
add(1, 2.0); // error, types not the same
add(1);       // error, size not greater than 1
```

The first call exemplified previously is correct, as there are two arguments, and both are of type `int`. The second call produces an error because the types of the arguments are different (`int` and `double`). Similarly, the third call also produces an error because only one argument was supplied.

The `HomogenousRange` concept can also be tested with the help of several `static_assert` statements, as shown next:

```
static_assert(HomogenousRange<int, int>);
static_assert(!HomogenousRange<int>);
static_assert(!HomogenousRange<int, double>);
```

We have walked through all the categories of the requires expressions that can be used for defining constraints. However, constraints can also be composed, and this is what we will discuss next.

Composing constraints

We have seen multiple examples of constraining template arguments but in all the cases so far, we used a single constraint. It is possible though for constraints to be composed using the && and || operators. A composition of two constraints using the && operator is called a **conjunction** and the composition of two constraints using the || operator is called a **disjunction**.

For a conjunction to be true, both constraints must be true. Like in the case of logical **AND** operations, the two constraints are evaluated from left to right, and if the left constraint is false, the right constraint is not evaluated. Let's look at an example:

```
template <typename T>
requires std::is_integral_v<T> && std::is_signed_v<T>
T decrement(T value)
{
    return value--;
}
```

In this snippet, we have a function template that returns the decremented value of the received argument. However, it only accepts signed integral values. This is specified with the conjunction of two constraints, std::is_integral_v<T> && std::is_signed_v<T>. The same result can be achieved using a different approach to defining the conjunction, as shown next:

```
template <typename T>
concept Integral = std::is_integral_v<T>;

template <typename T>
concept Signed = std::is_signed_v<T>;

template <typename T>
concept SignedIntegral = Integral<T> && Signed<T>;

template <SignedIngeral T>
T decrement(T value)
{
    return value--;
}
```

You can see three concepts defined here: one that constrains integral types, one that constrains signed types, and one that constrains integral and signed types.

Disjunctions work in a similar way. For a disjunction to be true, at least one of the constraints must be true. If the left constraint is true, then the right one is not evaluated. Again, let's see an example. If you recall the `add` function template from the first section of the chapter, we constrained it with the `std::is_arithmetic` type trait. However, we can get the same result using `std::is_integral` and `std::is_floating_point`, used as follows:

```
template <typename T>
requires std::is_integral_v<T> || std::is_floating_point_v<T>
T add(T a, T b)
{
    return a + b;
}
```

The expression `std::is_integral_v<T> || std::is_floating_point_v<T>` defines a disjunction of two atomic constraints. We will look at this kind of constraint in more detail later. For the time being, keep in mind that an atomic constraint is an expression of the `bool` type that cannot be decomposed into smaller parts. Similarly, to what we've done previously, we can also build a disjunction of concepts and use that. Here is how:

```
template <typename T>
concept Integral = std::is_integral_v<T>;

template <typename T>
concept FloatingPoint = std::is_floating_point_v<T>;

template <typename T>
concept Number = Integral<T> || FloatingPoint<T>;

template <Number T>
T add(T a, T b)
{
    return a + b;
}
```

As already mentioned, conjunctions and disjunctions are short-circuited. This has an important implication in checking the correctness of a program. Considering a conjunction of the form A<T> && B<T>, then A<T> is checked and evaluated first, and if it is false, the second constraint, B<T>, is not checked anymore.

Similarly, for the A<T> || B<T> disjunction, after A<T> is checked, if it evaluates to true, the second constraint, B<T>, will not be checked. If you want both conjunctions to be checked for well-formedness and then their Boolean value determined, then you must use the && and || operators differently. A conjunction or disjunction is formed only when the && and || tokens, respectively, appear nested in parentheses or as an operand of the && or || tokens. Otherwise, these operators are treated as logical operators. Let's explain this with examples:

```
template <typename T>
requires A<T> || B<T>
void f() {}


template <typename T>
requires (A<T> || B<T>)
void f() {}


template <typename T>
requires A<T> && (!A<T> || B<T>)
void f() {}
```

In all these examples, the || token defines a disjunction. However, when used inside a cast expression or a logical **NOT**, the && and || tokens define a logical expression:

```
template <typename T>
requires (!(A<T> || B<T>))
void f() {}


template <typename T>
requires (static_cast<bool>(A<T> || B<T>))
void f() {}
```

In these cases, the entire expression is first checked for correctness, and then its Boolean value is determined. It is worth mentioning that in this latter example both expressions, `!(A<T> || B<T>)` and `static_cast<bool>(A<T> || B<T>)`, need to be wrapped inside another set of parentheses because the expression of a requires clause cannot start with the `!` token or a cast.

Conjunctions and disjunctions cannot be used to constrain template parameter packs. However, there is a workaround to make it happen. Let's consider a variadic implementation of the add function template with the requirement that all arguments must be integral types. One would attempt to write such a constraint in the following form:

```
template <typename ... T>
requires std::is_integral_v<T> && ...
auto add(T ... args)
{
    return (args + ...);
}
```

This will generate a compiler error because the ellipsis is not allowed in this context. What we can do to avoid this error is to wrap the expression in a set of parentheses, as follows:

```
template <typename ... T>
requires (std::is_integral_v<T> && ...)
auto add(T ... args)
{
    return (args + ...);
}
```

The expression, `(std::is_integral_v<T> && ...)`, is now a fold expression. It is not a conjunction, as one would expect. Therefore, we get a single atomic constraint. The compiler will first check the correctness of the entire expression and then determine its Boolean value. To build a conjunction we first need to define a concept:

```
template <typename T>
concept Integral = std::is_integral_v<T>;
```

What we need to do next is change the requires clause so that it uses the newly defined concept and not the Boolean variable, `std::is_integral_v<T>`:

```
template <typename ... T>
requires (Integral<T> && ...)
auto add(T ... args)
{
    return (args + ...);
}
```

It does not look like much of a change but, in fact, because of the use of concepts, validating the correctness and determining the Boolean value occur individually for each template argument. If the constraint is not met for a type, the rest is short-circuited, and the validation stops.

You must have noticed that earlier in this section I used the term *atomic constraint* twice. Therefore, one would ask, what is an atomic constraint? It is an expression of the `bool` type that cannot be decomposed further. Atomic constraints are formed during the process of constraint normalization when the compiler decomposes constraints into conjunction and disjunctions of atomic constraints. This works as follows:

- The expression, `E1 && E2`, is decomposed into the conjunction of `E1` and `E2`.

- The expression, `E1 || E2`, is decomposed into the disjunction of `E1` and `E2`.

- The concept, `C<A1, A2, ... An>`, is replaced with its definition after substituting all the template arguments into its atomic constraints.

Atomic constraints are used for determining the partial ordering of constraints that, in turn, determine the partial ordering of function templates and class template specializations, as well as the next candidate for non-template functions in overload resolution. We will discuss this topic next.

Learning about the ordering of templates with constraints

When a compiler encounters function calls or class template instantiations, it needs to figure out what overload (for a function) or specialization (for a class) is the best match. A function may be overloaded with different type constraints. Class templates can also be specialized with different type constraints. In order to decide which is the best match, the compiler needs to figure out which one is the most constrained and, at the same time, evaluates to `true` after substituting all the template parameters. In order to figure this out, it performs the **constraints normalization**. This is the process of transforming the constraint expression into conjunctions and disjunctions of atomic constraints, as described at the end of the previous section.

An atomic constraint A is said to subsume another atomic constraint B if A implies B. A constraint declaration D1 whose constraints subsume the constraints of another declaration D2 is said to be at least as constrained as D2. Moreover, if D1 is at least as constrained as D2 but the reciprocal is not true, then it's said that D1 is more constrained than D2. More constrained overloads are selected as the best match.

We will discuss several examples in order to understand how constraints affect overload resolution. First, let's start with the following two overloads:

```
int add(int a, int b)
{
    return a + b;
}


template <typename T>
T add(T a, T b)
{
    return a + b;
}
```

The first overload is a non-template function that takes two `int` arguments and returns their sum. The second is the template implementation we have seen already in the chapter.

Having these two, let's consider the following calls:

```
add(1.0, 2.0);   // [1]
add(1, 2);       // [2]
```

The first call (at line [1]) takes two `double` values so only the template overload is a match. Therefore, its instantiation for the `double` type will be called. The second invocation of the add function (at line [2]) takes two integer arguments. Both overloads are a possible match. The compiler will select the most specific one, which is the non-template overload.

What if both overloads are templates but one of them is constrained? Here is an example to discuss:

```
template <typename T>
T add(T a, T b)
{
    return a + b;
}


template <typename T>
requires std::is_integral_v<T>
T add(T a, T b)
{
    return a + b;
}
```

The first overload is the function template seen previously. The second has an identical implementation except that it specifies a requirement for the template argument, which is restricted to integral types. If we consider the same two calls from the previous snippet, for the call at line [1] with two double values, only the first overload is a good match. For the call at line [2] with two integer values, both overloads are a good match. However, the second overload is more constrained (it has one constraint compared to the first one that has no constraint) so the compiler will select this one for the invocation.

In the next example, both overloads are constrained. The first overload requires that the size of the template argument is four, and the second overload requires that the template argument must be an integral type:

```
template <typename T>
requires (sizeof(T) == 4)
T add(T a, T b)
{
    return a + b;
}

template <typename T>
requires std::is_integral_v<T>
T add(T a, T b)
{
```

```
    return a + b;
}
```

Let's consider the following calls to this overloaded function template:

```
add((short)1, (short)2);   // [1]
add(1, 2);                 // [2]
```

The call at line [1] uses arguments of the short type. This is an integral type with the size 2; therefore, only the second overload is a match. However, the call at line [2] uses arguments of the int type. This is an integral type of size 4. Therefore, both overloads are a good match. However, this is an ambiguous situation, and the compiler is not able to select between the two and it will trigger an error.

What happens, though, if we change the two overloads slightly, as shown in the next snippet?

```
template <typename T>
requires std::is_integral_v<T>
T add(T a, T b)
{
    return a + b;
}


template <typename T>
requires std::is_integral_v<T> && (sizeof(T) == 4)
T add(T a, T b)
{
    return a + b;
}
```

Both overloads require that the template argument must be an integral type, but the second also requires that the size of the integral type must be 4 bytes. So, for the second overload, we use a conjunction of two atomic constraints. We will discuss the same two calls, with short arguments and with int arguments.

For the call at line [1], passing two short values, only the first overload is a good match, so this one will be invoked. For the call at line [2] that takes two int arguments, both overloads are a match. The second, however, is more constrained. Yet, the compiler is not able to decide which is a better match and will issue an ambiguous call error. This may be surprising to you because, in the beginning, I said that the most constrained overload will be selected from the overload set. It does not work in our example because we used type traits to constrain the two functions. The behavior is different if we instead use concepts. Here is how:

```
template <typename T>
concept Integral = std::is_integral_v<T>;


template <typename T>
requires Integral<T>
T add(T a, T b)
{
    return a + b;
}


template <typename T>
requires Integral<T> && (sizeof(T) == 4)
T add(T a, T b)
{
    return a + b;
}
```

There is no ambiguity anymore; the compiler will select the second overload as the best match from the overload set. This demonstrates that concepts are handled preferentially by the compiler. Remember, there are different ways to use constraints using concepts, but the preceding definition simply replaced a type trait with a concept; therefore, they are arguably a better choice for demonstrating this behavior than the next implementation:

```
template <Integral T>
T add(T a, T b)
{
    return a + b;
}


template <Integral T>
```

```
requires (sizeof(T) == 4)
T add(T a, T b)
{
    return a + b;
}
```

All the examples discussed in this chapter involved constraining function templates. However, it's possible to constrain non-template member functions as well as class templates and class template specializations. We will discuss these in the next sections, and we will start with the former.

Constraining non-template member functions

Non-template functions that are members of class templates can be constrained in a similar way to what we have seen so far. This enables template classes to define member functions only for types that satisfy some requirements. In the following example, the equality operator is constrained:

```
template <typename T>
struct wrapper
{
    T value;

    bool operator==(std::string_view str)
    requires std::is_convertible_v<T, std::string_view>
    {
        return value == str;
    }
};
```

The wrapper class holds a value of a T type and defines the operator== member only for types that are convertible to std::string_view. Let's see how this can be used:

```
wrapper<int>          a{ 42 };
wrapper<char const*> b{ "42" };


if(a == 42)   {} // error
if(b == "42") {} // OK
```

We have two instantiations of the `wrapper` class here, one for `int` and one for `char const*`. The attempt to compare the `a` object with the literal `42` generates a compiler error, because the `operator==` is not defined for this type. However, comparing the `b` object with the string literal `"42"` is possible because the equality operator is defined for types that can be implicitly converted to `std::string_view`, and `char const*` is such a type.

Constraining non-template members is useful because it's a cleaner solution than forcing members to be templates and using SFINAE. To understand this better let's consider the following implementation of the `wrapper` class:

```
template <typename T>
struct wrapper
{
    T value;

    wrapper(T const & v) :value(v) {}
};
```

This class template can be instantiated as follows:

```
wrapper<int> a = 42;               //OK

wrapper<std::unique_ptr<int>> p =
   std::make_unique<int>(42);    //error
```

The first line compiles successfully, but the second generates a compiler error. There are different messages issued by different compilers, but at the core of the error is the call to the implicitly deleted copy constructor of `std::unique_ptr`.

What we want to do is restrict the copy construction of `wrapper` from objects of the `T` type so that it only works for `T` types that are copy-constructible. The approach available before C++20 was to transform the copy constructor into a template and employ SFINAE. This would look as follows:

```
template <typename T>
struct wrapper
{
    T value;

    template <typename U,
```

```
        typename = std::enable_if_t<
                std::is_copy_constructible_v<U> &&
                std::is_convertible_v<U, T>>>
    wrapper(U const& v) :value(v) {}
};
```

This time we also get an error when trying to initialize a `wrapper<std::unique_ptr<int>>` from an `std::unique_ptr<int>` value but the errors are different. For instance, here are the error messages generated by Clang:

```
prog.cc:19:35: error: no viable conversion from 'typename __
unique_if<int>::__unique_single' (aka 'unique_ptr<int>') to
'wrapper<std::unique_ptr<int>>'
    wrapper<std::unique_ptr<int>> p = std::make_
unique<int>(42); // error
                                  ^
                                    ~~~~~~~~~~~~~~~~~~~~~~~~~~~


prog.cc:6:8: note: candidate constructor (the implicit copy
constructor) not viable: no known conversion from 'typename
__unique_if<int>::__unique_single' (aka 'unique_ptr<int>') to
'const wrapper<std::unique_ptr<int>> &' for 1st argument
struct wrapper
       ^


prog.cc:6:8: note: candidate constructor (the implicit move
constructor) not viable: no known conversion from 'typename
__unique_if<int>::__unique_single' (aka 'unique_ptr<int>') to
'wrapper<std::unique_ptr<int>> &&' for 1st argument
struct wrapper
       ^


prog.cc:13:9: note: candidate template ignored: requirement
'std::is_copy_constructible_v<std::unique_ptr<int,
std::default_delete<int>>>' was not satisfied [with U =
std::unique_ptr<int>]
        wrapper(U const& v) :value(v) {}
        ^
```

The most important message to help understand the cause of the problem is the last one. It says that the requirement that U substituted with `std::unique_ptr<int>` does not satisfy the Boolean condition. In C++20, we can do a better job at implementing the same restriction on the T template argument. This time, we can use constraints and the copy constructor does not need to be a template anymore. The implementation in C++20 can look as follows:

```
template <typename T>
struct wrapper
{
    T value;

    wrapper(T const& v)
        requires std::is_copy_constructible_v<T>
        :value(v)
    {}
};
```

Not only there is less code that does not require complicated SFINAE machinery, but it is simpler and easier to understand. It also generates potentially better error messages. In the case of Clang, the last note listed earlier is replaced with the following:

```
prog.cc:9:5: note: candidate constructor not viable:
constraints not satisfied
    wrapper(T const& v)
    ^

prog.cc:10:18: note: because 'std::is_copy_constructible_
v<std::unique_ptr<int> >' evaluated to false
        requires std::is_copy_constructible_v<T>
```

Before closing this section, it's worth mentioning that not only non-template functions that are members of classes can be constrained but also free functions. The use cases for non-template functions are rare and can be achieved with alternative simple solutions such as constexpr if. Let's look at an example, though:

```
void handle(int v)
{ /* do something */ }

void handle(long v)
```

```
    requires (sizeof(long) > sizeof(int))
{ /* do something else */ }
```

In this snippet, we have two overloads of the `handle` function. The first overload takes an `int` value and the second a `long` value. The body of these overloaded functions is not important but they should do different things, if and only if the size of `long` is different from the size of `int`. The standard specifies that the size of `int` is at least 16 bits, although on most platforms it is 32 bits. The size of `long` is at least 32 bits. However, there are platforms, such as **LP64**, where `int` is 32 bits and `long` is 64 bits. On these platforms, both overloads should be available. On all the other platforms, where the two types have the same size, only the first overload should be available. This can be defined in the form shown earlier, although the same can be achieved in C++17 with constexpr if as follows:

```
void handle(long v)
{
    if constexpr (sizeof(long) > sizeof(int))
    {
        /* do something else */
    }
    else
    {
        /* do something */
    }
}
```

In the next section, we'll learn how to use constraints to define restrictions on template arguments of class templates.

Constraining class templates

Class templates and class template specializations can also be constrained just like function templates. To start, we'll consider the `wrapper` class template again, but this time with the requirement that it should only work for template arguments of integral types. This can be simply specified in C++20 as follows:

```
template <std::integral T>
struct wrapper
{
    T value;
```

```
};


wrapper<int>    a{ 42 };    // OK
wrapper<double> b{ 42.0 };  // error
```

Instantiating the template for the `int` type is fine but does not work for `double` because this is not an integral type.

Requirements that also be specified with requires clauses and class template specializations can also be constrained. To demonstrate this, let's consider the scenario when we want to specialize the `wrapper` class template but only for types whose size is 4 bytes. This can be implemented as follows:

```
template <std::integral T>
struct wrapper
{
    T value;
};


template <std::integral T>
requires (sizeof(T) == 4)
struct wrapper<T>
{
    union
    {
        T value;
        struct
        {
            uint8_t byte4;
            uint8_t byte3;
            uint8_t byte2;
            uint8_t byte1;
        };
    };
};
```

We can use this class template as shown in the following snippet:

```
wrapper<short> a{ 42 };
std::cout << a.value << '\n';


wrapper<int> b{ 0x11223344 };
std::cout << std::hex << b.value << '\n';
std::cout << std::hex << (int)b.byte1 << '\n';
std::cout << std::hex << (int)b.byte2 << '\n';
std::cout << std::hex << (int)b.byte3 << '\n';
std::cout << std::hex << (int)b.byte4 << '\n';
```

The object a is an instance of `wrapper<short>`; therefore, the primary template is used. On the other hand, the object b is an instance of `wrapper<int>`. Since `int` has a size of 4 bytes (on most platforms) the specialization is used and we can access the individual types of the wrapped value through the `byte1`, `byte2`, `byte3`, and `byte4` members.

Lastly on this topic, we will discuss how variable templates and template aliases can also be constrained.

Constraining variable templates and template aliases

As you well know, apart from function templates and class templates we also have variable templates and alias templates in C++. These make no exception of the need to define constraints. The same rules for constraining the template arguments discussed so far apply to these two. In this section, we will demonstrate them shortly. Let's start with variable templates.

It is a typical example to define the `PI` constant for showing how variable templates work. Indeed, it is a simple definition that looks as follows:

```
template <typename T>
constexpr T PI = T(3.14159265358979932385L);
```

However, this only makes sense for floating-point types (and maybe other types such as decimal, which does not exist in C++ yet). Therefore, this definition should be restricted to floating-point types, as follows:

```
template <std::floating_point T>
constexpr T PI = T(3.1415926535897932385L);


std::cout << PI<double> << '\n';   // OK
std::cout << PI<int> << '\n';      // error
```

The use of PI<double> is correct but PI<int> produces a compiler error. This is what constraints can provide in a simple and readable manner.

Finally, the last category of templates that we have in the language, alias templates, can also be constrained. In the following snippet, we can see such an example:

```
template <std::integral T>
using integral_vector = std::vector<T>;
```

The integral_vector template is an alias for std::vector<T> when T is an integral type. The very same can be achieved with the following alternative, although longer, declaration:

```
template <typename T>
requires std::integral<T>
using integral_vector = std::vector<T>;
```

We can use this integral_vector alias template as follows:

```
integral_vector<int>    v1 { 1,2,3 };        // OK
integral_vector<double> v2 {1.0, 2.0, 3.0}; // error
```

Defining the v1 object works fine since int is an integral type. However, defining the v2 vector generates a compiler error because double is not an integral type.

If you paid attention to the examples in this section, you will have noticed that they don't use the type traits (and the associated variable templates) we used previously in the chapter, but a couple of concepts: std::integral and std::floating_point. These are defined in the <concepts> header and help us avoid repeatedly defining the same concepts based on available C++11 (or newer) type traits. We will look at the content of the standard concepts library shortly. Before we do that, let's see what other ways we can employ to define constraints in C++20.

Learning more ways to specify constraints

We have discussed in this chapter about requires clauses and requires expressions. Although both are introduced with the new `requires` keyword, they are different things and should be fully understood:

- A *requires clause* determines whether a function participates in overload resolution or not. This happens based on the value of a compile-time Boolean expression.

- A *requires expression* determines whether a set of one or more expressions is well-formed, without having any side effects on the behavior of the program. A requires expression is a Boolean expression that can be used with a requires clause.

Let's see an example again:

```
template <typename T>
concept addable = requires(T a, T b) { a + b; };
                            // [1] requires expression


template <typename T>
requires addable<T>    // [2] requires clause
auto add(T a, T b)
{
    return a + b;
}
```

The construct at line [1] that starts with the `requires` keyword is a requires expression. It verifies that the expression, a + b, is well-formed for any T. On the other hand, the construct at line [2] is a requires clause. If the Boolean expression `addable<T>` evaluates to `true`, the function takes part in overload resolution; otherwise, it does not.

Although requires clauses are supposed to use concepts, a requires expression can also be used. Basically, anything that can be placed on the right-hand side of the = token in a concept definition can be used with a requires clause. That means we can do the following:

```
template <typename T>
    requires requires(T a, T b) { a + b; }
auto add(T a, T b)
{
    return a + b;
}
```

Although this is perfectly legal code it is arguable whether it's a good way of using constraints. I would recommend avoiding creating constructs that start with `requires requires`. They are less readable and may create confusion. Moreover, named concepts can be used anywhere, while a requires clause with a requires expression will have to be duplicated if it needs to be used for multiple functions.

Now that we've seen how to constrain template arguments in several ways using constraints and concepts, let's see how we can simplify function template syntax and constrain the template arguments.

Using concepts to constrain auto parameters

In *Chapter 2, Template Fundamentals*, we discussed generic lambdas, introduced in C++14, as well as lambda templates, introduced in C++20. A lambda that uses the `auto` specifier for at least one parameter is called a **generic lambda**. The function object generated by the compiler will have a templated call operator. Here is an example to refresh your memory:

```
auto lsum = [](auto a, auto b) {return a + b; };
```

The C++20 standard generalizes this feature for all functions. You can use the `auto` specifier in the function parameter list. This has the effect of transforming the function into a template function. Here is an example:

```
auto add(auto a, auto b)
{
    return a + b;
}
```

This is a function that takes two parameters and returns their sum (or to be more precise, the result of applying `operator+` on the two values). Such a function using `auto` for function parameters is called an **abbreviated function template**. It is basically shorthand syntax for a function template. The equivalent template for the previous function is the following:

```
template<typename T, typename U>
auto add(T a, U b)
{
    return a + b;
}
```

We can call this function as we would call any template function, and the compiler will generate the proper instantiations by substituting the template arguments with the actual types. For instance, let's consider the following calls:

```
add(4, 2);    // returns 6
add(4.0, 2); // returns 6.0
```

We can use the cppinsights.io website to check the compiler-generated code for the add abbreviated function template based on these two calls. The following specializations are generated:

```
template<>
int add<int, int>(int a, int b)
{
    return a + b;
}

template<>
double add<double, int>(double a, int b)
{
    return a + static_cast<double>(b);
}
```

Since an abbreviated function template is nothing but a regular function template with a simplified syntax, such a function can be explicitly specialized by the user. Here is an example:

```
template<>
auto add(char const* a, char const* b)
{
    return std::string(a) + std::string(b);
}
```

This is a full specialization for the char const* type. This specialization enables us to make calls such as add("4", "2"), although the result is a std::string value.

This category of abbreviated function templates is called **unconstrained**. There is no restriction on the template arguments. However, it is possible to provide constraints for their parameters with concepts. Abbreviated function templates that use concepts are called **constrained**. Next, you can see an example of an add function constrained for integral types:

```
auto add(std::integral auto a, std::integral auto b)
{
    return a + b;
}
```

If we consider again the same calls we saw earlier, the first would be successful, but the second would produce a compiler error because there is no overload that takes a double and an int value:

```
add(4, 2);    // OK
add(4.2, 0); // error
```

Constrained auto can also be used for variadic abbreviated function templates. An example is shown in the following snippet:

```
auto add(std::integral auto ... args)
{
    return (args + ...);
}
```

Last but not least, constrained auto can be used with generic lambdas too. If we would like the generic lambda shown at the beginning of this section to be used only with integral types, then we can constrain it as follows:

```
auto lsum = [](std::integral auto a, std::integral auto b)
{
    return a + b;
};
```

With the closing of this section, we have seen all the language features related to concepts and constraints in C++20. What is left to discuss is the set of concepts provided by the standard library, of which we have seen a couple already. We will do this next.

Exploring the standard concepts library

The standard library provides a set of fundamental concepts that can be used to define requirements on the template arguments of function templates, class templates, variable templates, and alias templates, as we have seen throughout this chapter. The standard concepts in C++20 are spread across several headers and namespaces. We will present some of them in this section although not all of them. You can find all of them online at `https://en.cppreference.com/`.

The main set of concepts is available in the `<concepts>` header and the `std` namespace. Most of these concepts are equivalent to one or more existing type traits. For some of them, their implementation is well-defined; for some, it is unspecified. They are grouped into four categories: core language concepts, comparison concepts, object concepts, and callable concepts. This set of concepts contains the following (but not only):

Concept	Description
`same_as`	Defines the requirement that a type T is the same as another type U.
`derived_from`	Defines the requirement that a type D is derived from another type B.
`convertible_to`	Defines the requirement that a type T is implicitly convertible to another type U.
`common_reference_with`	Defines the requirements that two types, T and U, have a common reference type.
`common_with`	Defines the requirement that two types, T and U, have a common type to which both can be convertible.
`integral`	Defines the requirement that a type T is an integral type.
`signed_integral`	Defines the requirement that a type T is a signed integral type.
`unsigned_integral`	Defines the requirement that a type T is an unsigned integral type.
`floating_point`	Defines the requirement that a type T is a floating-point type.
`assignable_from`	Defines the requirement that an expression of a type U can be assigned to an lvalue expression of a type T.

Concept	Description
swappable	Defines the requirement that two values of the same type T can be swapped.
swappable_with	Defines the requirement that a value of a type T can be swapped with a value of a type U.
destructible	Defines the requirement that a value of a type T can safely be destroyed (without any exception thrown from the destructor).
constructible_from	Defines the requirement that an object of a type T can be constructed with the given set of argument types.
default_initializable	Defines the requirement that an object of a type T can be default-constructible (either value initialized T(), direct-list-initialized from an empty initializer list T{}, or default-initialized, as in T t;).
move_constructible	Defines the requirement that an object of a type T can be constructed with move semantics.
copy_constructible	Defines the requirement that an object of a type T can be copy constructed and move constructed.
moveable	Defines the requirement that an object of a type T can be moved and swapped.
copyable	Defines the requirement that an object of a type T can be copied, moved, and swapped.
regular	Defines the requirement that a type T satisfies both the semiregular and equality_comparable concepts.
semiregular	Defines the requirement that an object of a type T can be copied, moved, swapped, and default constructed.
equality_comparable	Defines the requirement that the comparison operator == for a type T reflects equality, meaning that it yields true if and only if two values are equal. Similarly, the =! comparison reflects inequality.
predicate	Defines the requirement that a callable type T is a Boolean predicate.

Table 6.1

Some of these concepts are defined using type traits, some are a combination of other concepts or concepts and type traits, and some have, at least partially, an unspecified implementation. Here are some examples:

```
template < class T >
concept integral = std::is_integral_v<T>;

template < class T >
concept signed_integral = std::integral<T> &&
                          std::is_signed_v<T>;

template <class T>
concept regular = std::semiregular<T> &&
                  std::equality_comparable<T>;
```

C++20 also introduces a new system of iterators, based on concepts, and defines a set of concepts in the `<iterator>` header. Some of these concepts are listed in the following table:

Concept	Description
indirectly_readable	Defines the requirement that values of a type can be read by applying the * operator.
indirectly_writable	Defines the requirement that objects referenced by an iterator type can be written to.
input_iterator	Defines the requirement that a type is an input iterator (supports reading, pre-, and post-increment).
output_iterator	Defines the requirement that a type is an output iterator (supports writing, pre-, and post-increment).
forward_iterator	Defines the requirement that a type that is an input_iterator is also a forward iterator (supports equality comparison and multi-pass).
bidirectional_iterator	Defines the requirement that a type that is a forward_iterator is also a bidirectional iterator (supports moving backward).

Concept	Description
`random_access_iterator`	Defines the requirement that a type that is a `bidirectional_iterator` is also a random-access iterator (supports subscripting and advancement in constant time).
`contiguous_iterator`	Defines the requirement that a type that is a `random_access_iterator` is also a contiguous iterator (elements are stored in contiguous memory locations).

Table 6.2

Here is how the `random_access_iterator` concept is defined in the C++ standard:

```
template<typename I>
concept random_access_iterator =
    std::bidirectional_iterator<I> &&
    std::derived_from</*ITER_CONCEPT*/<I>,
                      std::random_access_iterator_tag> &&
    std::totally_ordered<I> &&
    std::sized_sentinel_for<I, I> &&
    requires(I i,
             const I j,
             const std::iter_difference_t<I> n)
    {
        { i += n } -> std::same_as<I&>;
        { j +  n } -> std::same_as<I>;
        { n +  j } -> std::same_as<I>;
        { i -= n } -> std::same_as<I&>;
        { j -  n } -> std::same_as<I>;
        {   j[n]  } -> std::same_as<std::iter_reference_t<I>>;
    };
```

As you can see, it uses several concepts (some of them not listed here) as well as a requires expression to ensure that some expressions are well-formed.

Also, in the `<iterator>` header, there is a set of concepts designed to simplify the constraining of general-purpose algorithms. Some of these concepts are listed in the next table:

Concept	Description
`indirectly_movable`	Defines the requirement that values can be moved from an `indirectly_readable` type to an `indirectly_readable` type.
`indirectly_copyable`	Defines the requirement that values can be copied from an `indirectly_readable` type to an `indirectly_copyable` type.
`mergeable`	Defines the requirements for algorithms that merge sorted sequences into an output sequence by copying elements.
`sortable`	Defines the requirements for algorithms that modify sequences into ordered sequences.
`permutable`	Defines the requirements for algorithms that reorder elements in place.

Table 6.3

One of the several major features included in C++20 (along with concepts, modules, and coroutines) are ranges. The `ranges` library defines a series of classes and functions for simplifying operations with ranges. Among these is a set of concepts. These are defined in the `<ranges>` header and the `std::ranges` namespace. Some of these concepts are listed as follows:

Concept	Description
`range`	Defines the requirement that a type R is a range, meaning that it provides a begin iterator and an end sentinel.
`sized_range`	Defines the requirement that a type R is a range with the size known in constant time.
`view`	Defines the requirement that a type R is a view, meaning that it provides constant-time copy, move, and assignment operations.
`input_range`	Defines the requirement that a type R is a range whose iterator type satisfies the `input_iterator` concept.
`output_range`	Defines the requirement that a type R is a range whose iterator type satisfies the `output_iterator` concept.
`forward_range`	Defines the requirement that a type R is a range whose iterator type satisfies the `forward_iterator` concept.

Concept	Description
`bidirectional_range`	Defines the requirement that a type R is a range whose iterator type satisfies the `bidirectional_iterator` concept.
`random_access_range`	Defines the requirement that a type R is a range whose iterator type satisfies the `random_access_iterator` concept.
`contiguous_range`	Defines the requirement that a type R is a range whose iterator type satisfies the `contiguous_iterator` concept.

Table 6.4

Here is how some of these concepts are defined:

```
template< class T >
concept range = requires ( T& t ) {
    ranges::begin(t);
    ranges::end  (t);
};


template< class T >
concept sized_range = ranges::range<T> &&
    requires (T& t) {
        ranges::size(t);
    };


template< class T >
concept input_range = ranges::range<T> &&
    std::input_iterator<ranges::iterator_t<T>>;
```

As mentioned already, there are more concepts than those listed here. Others will probably be added in the future. This section is not intended as a complete reference to the standard concepts but rather as an introduction to them. You can learn more about each of these concepts from the official C++ reference documentation available at `https://en.cppreference.com/`. As for ranges, we will learn more about them and explore what the standard library provides in *Chapter 8, Ranges and Algorithms*.

Summary

The C++20 standard introduced some new major features to the language and the standard library. One of these is concepts, which was the topic of this chapter. A concept is a named constraint that can be used to define requirements on template arguments for function templates, class templates, variable templates, and alias templates.

In this chapter, we have explored in detail how we can use constraints and concepts and how they work. We have learned about requires clauses (that determine whether a template participates in overload resolution) and requires expressions (that specify requirements for well-formedness of expressions). We have seen what various syntaxes are for specifying constraints. We also learned about abbreviated function templates that provide a simplified syntax for function templates. At the end of the chapter, we explored the fundamental concepts available in the standard library.

In the next chapter, we will shift our attention toward applying the knowledge accumulated so far to implement various template-based patterns and idioms.

Questions

1. What are constraints and what are concepts?

2. What are a requires clause and a requires expression?

3. What are the categories of requires expressions?

4. How do constraints affect the ordering of templates in overload resolution?

5. What are abbreviated function templates?

Further reading

- *C++20 Concepts - A Quick Introduction*, Bartlomiej Filipek, `https://www.cppstories.com/2021/concepts-intro/`

- *How C++20 Concepts can simplify your code*, Andreas Fertig, `https://andreasfertig.blog/2020/07/how-cpp20-concepts-can-simplify-your-code/`

- *What are C++20 concepts and constraints? How to use them?*, Sorush Khajepor, `https://iamsorush.com/posts/concepts-cpp/`

- *Requires-clause*, Andrzej Krzemieński, `https://akrzemi1.wordpress.com/2020/03/26/requires-clause/`

- *Ordering by constraints*, Andrzej Krzemieński, `https://akrzemi1.wordpress.com/2020/05/07/ordering-by-constraints/`

Part 3: Applied Templates

In this part, you will put to practice the knowledge of templates you have accumulated so far. You will learn about static polymorphism and patterns such as the Curiously Recuring Template Pattern and mixins, as well as type erasure, tag dispatching, expression templates, and typelists. You will also learn about the design of standard containers, iterators, and algorithms and you will learn to implement your own. We will explore the C++20 Ranges library with its ranges and constrained algorithms and you will learn how to write your own range adaptor.

This section comprises the following chapters:

- *Chapter 7, Patterns and Idioms*
- *Chapter 8, Ranges and Algorithms*
- *Chapter 9, The Ranges Library*

7
Patterns and Idioms

The previous parts of the book were designed to help you learn everything about templates, from the basics to the most advanced features, including the latest concepts and constraints from C++20. Now, it is time for us to put this knowledge to work and learn about various metaprogramming techniques. In this chapter, we will discuss the following topics:

- Dynamic versus static polymorphism
- The **Curiously Recurring Template Pattern (CRTP)**
- Mixins
- Type erasure
- Tag dispatching
- Expression templates
- Typelists

By the end of the chapter, you will have a good understanding of various multiprogramming techniques that will help you solve a variety of problems.

Let's start the chapter by discussing the two forms of polymorphism: dynamic and static.

Dynamic versus static polymorphism

When you learn about object-oriented programming, you learn about its fundamental principles, which are **abstraction**, **encapsulation**, **inheritance**, and **polymorphism**. C++ is a multi-paradigm programming language that supports object-oriented programming too. Although a broader discussion on the principles of object-oriented programming is beyond the scope of this chapter and this book, it is worth discussing at least some aspects related to polymorphism.

So, what is polymorphism? The term is derived from the Greek words for *"many forms"*. In programming, it's the ability of objects of different types to be treated as if they were of the same type. The C++ standard actually defines a polymorphic class as follows (see C++20 standard, paragraph *11.7.2, Virtual functions*):

> *A class that declares or inherits a virtual function is called a polymorphic class.*

It also defines polymorphic objects based on this definition, as follows (see C++20 standard, paragraph *6.7.2, Object model*):

> *Some objects are polymorphic (11.7.2); the implementation generates information associated with each such object that makes it possible to determine that object's type during program execution.*

However, this actually refers to what is called **dynamic polymorphism** (or late binding), but there is yet another form of polymorphism, called **static polymorphism** (or early binding). Dynamic polymorphism occurs at runtime with the help of interfaces and virtual functions, while static polymorphism occurs at compile-time with the help of overloaded functions and templates. This is described in Bjarne Stroustrup's glossary of terms for the C++ language (see https://www.stroustrup.com/glossary.html):

> *polymorphism - providing a single interface to entities of different types. virtual functions provide dynamic (run-time) polymorphism through an interface provided by a base class. Overloaded functions and templates provide static (compile-time) polymorphism.*

Let's look at an example of dynamic polymorphism. The following is a hierarchy of classes representing different units in a game. These units may attack others, so there is a base class with a pure virtual function called attack, and several derived classes implementing specific units that override this virtual function doing different things (of course, for simplicity, here we just print a message to the console). It looks as follows:

```cpp
struct game_unit
{
    virtual void attack() = 0;
};

struct knight : game_unit
{
    void attack() override
    { std::cout << "draw sword\n"; }
};

struct mage : game_unit
{
    void attack() override
    { std::cout << "spell magic curse\n"; }
};
```

Based on this hierarchy of classes (which according to the standard are called **polymorphic classes**), we can write the function fight shown as follows. This takes a sequence of pointers to objects of the base game_unit type and calls the attack member function. Here is its implementation:

```cpp
void fight(std::vector<game_unit*> const & units)
{
    for (auto unit : units)
    {
        unit->attack();
    }
}
```

This function does not need to know the actual type of each object because due to dynamic polymorphism, it can handle them as if they were of the same (base) type. Here is an example of using it:

```
knight k;
mage m;
fight({&k, &m});
```

But now let's say you could combine a mage and a knight and create a new unit, a knight mage with special abilities from both these units. C++ enables us to write code as follows:

```
knight_mage km = k + m;
km.attack();
```

This does not come out of the box, but the language supports overloading operators, and we could do that for any user-defined types. To make the preceding line possible, we need the following:

```
struct knight_mage : game_unit
{
    void attack() override
    { std::cout << "draw magic sword\n"; }
};

knight_mage operator+(knight const& k, mage const& m)
{
    return knight_mage{};
}
```

Keep in mind these are just some simple snippets without any complex code. But the ability to add a `knight` and a `mage` together to create a `knight_mage` is nothing short of the ability to add two integers together, or a `double` and an `int`, or two `std::string` objects. This happens because there are many overloads of the + operator (both for built-in types and user-defined types) and based on the operands, the compiler is selecting the appropriate overload. Therefore, it can be said there are many forms of this operator. This is true for all the operators that can be overloaded; the + operator is just a typical example since it is ubiquitous. And this is the compile-time version of polymorphism, called **static polymorphism**.

Operators are not the only functions that can be overloaded. Any function can be overloaded. Although we have seen many examples in the book, let's take another one:

```
struct attack   { int value; };
struct defense { int value; };


void increment(attack& a)   { a.value++; }
void increment(defense& d) { d.value++; }
```

In this snippet, the `increment` function is overloaded for both the `attack` and `defense` types, allowing us to write code as follows:

```
attack a{ 42 };
defense d{ 50 };


increment(a);
increment(d);
```

We can replace the two overloads of `increment` with a function template. The changes are minimal, as shown in the next snippet:

```
template <typename T>
void increment(T& t) { t.value++; }
```

The previous code continues to work, but there is a significant difference: in the former example, we had two overloads, one for `attack` and one for `defense`, so you could call the function with objects of these types but nothing else. In the latter, we have a template that defines a family of overloaded functions for any possible type `T` that has a data member called `value` whose type supports the post-increment operator. We can define constraints for such a function template, which is something we have seen in the previous two chapters of the book. However, the key takeaway is that overloaded functions and templates are the mechanisms to implement static polymorphism in the C++ language.

Dynamic polymorphism incurs a performance cost because in order to know what functions to call, the compiler needs to build a table of pointers to virtual functions (and also a table of pointers to virtual base classes in case of virtual inheritance). So, there is some level of indirection when calling virtual functions polymorphically. Moreover, the details of virtual functions are not made available to the compiler who cannot optimize them.

When these things can be validated as performance issues, we could raise the question: can we get the benefits of dynamic polymorphism at compile time? The answer is yes and there is one way to achieve this: the Curiously Recurring Template Pattern, which we will discuss next.

The Curiously Recurring Template Pattern

This pattern has a rather curious name: the **Curiously Recurring Template Pattern**, or **CRTP** for short. It's called curious because it is rather odd and unintuitive. The pattern was first described (and its name coined) by James Coplien in a column in the *C++ Report* in 1995. This pattern is as follows:

- There is a base class template that defines the (static) interface.

- Derived classes are themselves the template argument for the base class template.

- The member functions of the base class call member functions of its type template parameter (which are the derived classes).

Let's see how the pattern implementation looks in practice. We will transform the previous example with game units into a version using the CRTP. The pattern implementation goes as follows:

```cpp
template <typename T>
struct game_unit
{
    void attack()
    {
        static_cast<T*>(this)->do_attack();
    }
};

struct knight : game_unit<knight>
{
    void do_attack()
    { std::cout << "draw sword\n"; }
};

struct mage : game_unit<mage>
{
```

```
    void do_attack()
    { std::cout << "spell magic curse\n"; }
};
```

The game_unit class is now a template class but contains the same member function, attack. Internally, this performs an upcast of the this pointer to T* and then invokes a member function called do_attack. The knight and mage classes derive from the game_unit class and pass themselves as the argument for the type template parameter T. Both provide a member function called do_attack.

Notice that the member function in the base class template and the called member function in the derived classes have different names. Otherwise, if they had the same name, the derived class member functions would hide the member from the base since these are no longer virtual functions.

The fight function that takes a collection of game units and calls the attack function needs to change too. It needs to be implemented as a function template, as follows:

```
template <typename T>
void fight(std::vector<game_unit<T>*> const & units)
{
    for (auto unit : units)
    {
        unit->attack();
    }
}
```

Using this function is a little different than before. It goes as follows:

```
knight k;
mage    m;
fight<knight>({ &k });
fight<mage>({ &m });
```

We have moved the runtime polymorphism to compile-time. Therefore, the `fight` function cannot treat `knight` and `mage` objects polymorphically. Instead, we get two different overloads, one that can handle `knight` objects and one that can handle `mage` objects. This is static polymorphism.

Although the pattern might not look complicated after all, the question you're probably asking yourself at this point is: how is this pattern actually useful? There are different problems you can solve using CRT, including the following:

- Limiting the number of times a type can be instantiated
- Adding common functionality and avoiding code duplication
- Implementing the composite design pattern

In the following subsections, we will look at each of these problems and see how to solve them with CRTP.

Limiting the object count with CRTP

Let's assume that for the game in which we created knights and mages we need some items to be available in a limited number of instances. For instance, there is a special sword type called *Excalibur* and there should be only one instance of it. On the other hand, there is a book of magic spells but there cannot be more than three instances of it at a time in the game. How do we solve this? Obviously, the sword problem could be solved with the singleton pattern. But what do we do when we need to limit the number to some higher value but still finite? The singleton pattern wouldn't be of much help (unless we transform it into a *"multiton"*) but the CRTP would.

First, we start with a base class template. The only thing this class template does is keep a count of how many times it has been instantiated. The counter, which is a static data member, is incremented in the constructor and decremented in the destructor. When that count exceeds a defined limit, an exception is thrown. Here is the implementation:

```
template <typename T, size_t N>
struct limited_instances
{
    static std::atomic<size_t> count;
    limited_instances()
    {
        if (count >= N)
            throw std::logic_error{ "Too many instances" };
        ++count;
```

```
    }
    ~limited_instances() { --count; }
};
```

```
template <typename T, size_t N>
std::atomic<size_t> limited_instances<T, N>::count = 0;
```

The second part of the template consists of defining the derived classes. For the mentioned problem, they are as follows:

```
struct excalibur : limited_instances<excalibur, 1>
{};
```

```
struct book_of_magic : limited_instances<book_of_magic, 3>
{};
```

We can instantiate `excalibur` once. The second time we try to do the same (while the first instance is still alive) an exception will be thrown:

```
excalibur e1;
try
{
    excalibur e2;
}
catch (std::exception& e)
{
    std::cout << e.what() << '\n';
}
```

Similarly, we can instantiate `book_of_magic` three times and an exception will be thrown the fourth time we attempt to do that:

```
book_of_magic b1;
book_of_magic b2;
book_of_magic b3;
try
{
    book_of_magic b4;
}
```

```
catch (std::exception& e)
{
    std::cout << e.what() << '\n';
}
```

Next, we look at a more common scenario, adding common functionality to types.

Adding functionality with CRTP

Another case when the curiously recurring template pattern can help us is providing common functionalities to derived classes through generic functions in a base class that relies solely on derived class members. Let's take an example to understand this use case.

Let's suppose that some of our game units have member functions such as `step_forth` and `step_back` that move them one position, forward or backward. These classes would look as follows (at a bare minimum):

```
struct knight
{
    void step_forth();
    void step_back();
};

struct mage
{
    void step_forth();
    void step_back();
};
```

However, it could be a requirement that everything that can move back and forth one step should also be able to advance or retreat an arbitrary number of steps. However, this functionality could be implemented based on the `step_forth` and `step_back` functions, which would help avoid having duplicate code in each of these game unit classes. The CRTP implementation for this problem would, therefore, look as follows:

```
template <typename T>
struct movable_unit
{
    void advance(size_t steps)
    {
```

```
        while (steps--)
            static_cast<T*>(this)->step_forth();
    }

    void retreat(size_t steps)
    {
        while (steps--)
            static_cast<T*>(this)->step_back();
    }
};

struct knight : movable_unit<knight>
{
    void step_forth()
    { std::cout << "knight moves forward\n"; }

    void step_back()
    { std::cout << "knight moves back\n"; }
};

struct mage : movable_unit<mage>
{
    void step_forth()
    { std::cout << "mage moves forward\n"; }

    void step_back()
    { std::cout << "mage moves back\n"; }
};
```

We can advance and retreat the units by calling the base class `advance` and `retreat` member functions as follows:

```
knight k;
k.advance(3);
k.retreat(2);
```

```
mage m;
m.advance(5);
m.retreat(3);
```

You could argue that the same result could be achieved using non-member function templates. For the sake of discussion, such a solution is presented as follows:

```
struct knight
{
    void step_forth()
    { std::cout << "knight moves forward\n"; }

    void step_back()
    { std::cout << "knight moves back\n"; }
};

struct mage
{
    void step_forth()
    { std::cout << "mage moves forward\n"; }

    void step_back()
    { std::cout << "mage moves back\n"; }
};

template <typename T>
void advance(T& t, size_t steps)
{
    while (steps--) t.step_forth();
}

template <typename T>
void retreat(T& t, size_t steps)
{
    while (steps--) t.step_back();
}
```

The client code would need to change but the changes are actually small:

```
knight k;
advance(k, 3);
retreat(k, 2);


mage m;
advance(m, 5);
retreat(m, 3);
```

The choice between these two may depend on the nature of the problem and your preferences. However, the CRTP has the advantage that it is describing well the interface of the derived classes (such as knight and mage in our example). With the non-member functions, you wouldn't necessarily know about this functionality, which would probably come from a header that you need to include. However, with CRTP, the class interface is well visible to those using it.

For the last scenario we discuss here, let's see how CRTP helps to implement the composite design pattern.

Implementing the composite design pattern

In their famous book, *Design Patterns: Elements of Reusable Object-Oriented Software*, the Gang of Four (Erich Gamma, Richard Helm, Ralph Johnson, and John Vlissides) describe a structural pattern called composite that enables us to compose objects into larger structures and treat both individual objects and compositions uniformly. This pattern can be used when you want to represent part-whole hierarchies of objects and you want to ignore the differences between individual objects and compositions of individual objects.

To put this pattern into practice, let's consider the game scenario again. We have heroes that have special abilities and can do different actions, one of which is allying with another hero. That can be easily modeled as follows:

```
struct hero
{
    hero(std::string_view n) : name(n) {}

    void ally_with(hero& u)
    {
        connections.insert(&u);
        u.connections.insert(this);
```

```
    }
private:
    std::string name;
    std::set<hero*> connections;

    friend std::ostream& operator<<(std::ostream& os,
                                    hero const& obj);
};

std::ostream& operator<<(std::ostream& os,
                         hero const& obj)
{
    for (hero* u : obj.connections)
        os << obj.name << " --> [" << u->name << "]" << '\n';

    return os;
}
```

These heroes are represented by the hero class that contains a name, a list of connections to other hero objects, as well as a member function, ally_with, that defines an alliance between two heroes. We can use it as follows:

```
hero k1("Arthur");
hero k2("Sir Lancelot");
hero k3("Sir Gawain");

k1.ally_with(k2);
k2.ally_with(k3);

std::cout << k1 << '\n';
std::cout << k2 << '\n';
std::cout << k3 << '\n';
```

The output of running this snippet is the following:

```
Arthur --> [Sir Lancelot]


Sir Lancelot --> [Arthur]
Sir Lancelot --> [Sir Gawain]


Sir Gawain --> [Sir Lancelot]
```

Everything was simple so far. But the requirement is that heroes could be grouped together to form parties. It should be possible for a hero to ally with a group, and for a group to ally with either a hero or an entire group. Suddenly, there is an explosion of functions that we need to provide:

```
struct hero_party;

struct hero
{
    void ally_with(hero& u);
    void ally_with(hero_party& p);
};

struct hero_party : std::vector<hero>
{
    void ally_with(hero& u);
    void ally_with(hero_party& p);
};
```

This is where the composite design pattern helps us treat heroes and parties uniformly and avoid unnecessary duplications of the code. As usual, there are different ways to implement it, but one way is using the curiously recurring template pattern. The implementation requires a base class that defines the common interface. In our case, this will be a class template with a single member function called ally_with:

```
template <typename T>
struct base_unit
{
```

```
    template <typename U>
    void ally_with(U& other);
};
```

We will define the `hero` class as a derived class from `base_unit<hero>`. This time, the `hero` class no longer implements `ally_with` itself. However, it features `begin` and `end` methods that are intended to simulate the behavior of a container:

```
struct hero : base_unit<hero>
{
    hero(std::string_view n) : name(n) {}

    hero* begin() { return this; }
    hero* end() { return this + 1; }

private:
    std::string name;
    std::set<hero*> connections;

    template <typename U>
    friend struct base_unit;

    template <typename U>
    friend std::ostream& operator<<(std::ostream& os,
                                    base_unit<U>& object);
};
```

The class that models a group of heroes is called `hero_party` and derives from both `std::vector<hero>` (to define a container of `hero` objects) and from `base_unit<hero_party>`. This is why the `hero` class has `begin` and `end` functions to help us perform iterating operations on `hero` objects, just as we would do with `hero_party` objects:

```
struct hero_party : std::vector<hero>,
                    base_unit<hero_party>
{};
```

We need to implement the `ally_with` member function of the base class. The code is shown as follows. What it does is iterate through all the sub-objects of the current object and connect them with all the sub-objects of the supplied argument:

```
template <typename T>
template <typename U>
void base_unit<T>::ally_with(U& other)
{
    for (hero& from : *static_cast<T*>(this))
    {
        for (hero& to : other)
        {
            from.connections.insert(&to);
            to.connections.insert(&from);
        }
    }
}
```

The `hero` class declared the `base_unit` class template a friend so that it could access the `connections` member. It also declared the `operator<<` as a friend so that this function could access both the `connections` and `name` private members. For more information about templates and friends, see *Chapter 4, Advanced Template Concepts*. The output stream operator implementation is shown here:

```
template <typename T>
std::ostream& operator<<(std::ostream& os,
                         base_unit<T>& object)
{
    for (hero& obj : *static_cast<T*>(&object))
    {
        for (hero* n : obj.connections)
            os << obj.name << " --> [" << n->name << "]"
               << '\n';
    }
    return os;
}
```

Having all this defined, we can write code as follows:

```
hero k1("Arthur");
hero k2("Sir Lancelot");

hero_party p1;
p1.emplace_back("Bors");

hero_party p2;
p2.emplace_back("Cador");
p2.emplace_back("Constantine");

k1.ally_with(k2);
k1.ally_with(p1);

p1.ally_with(k2);
p1.ally_with(p2);

std::cout << k1 << '\n';
std::cout << k2 << '\n';
std::cout << p1 << '\n';
std::cout << p2 << '\n';
```

We can see from this that we are able to ally a `hero` with both another `hero` and a `hero_party`, as well as a `hero_party` with either a `hero` or another `hero_party`. That was the proposed goal, and we were able to do it without duplicating the code between `hero` and `hero_party`. The output of executing the previous snippet is the following:

```
Arthur --> [Sir Lancelot]
Arthur --> [Bors]

Sir Lancelot --> [Arthur]
Sir Lancelot --> [Bors]

Bors --> [Arthur]
Bors --> [Sir Lancelot]
Bors --> [Cador]
```

```
Bors --> [Constantine]

Cador --> [Bors]
Constantine --> [Bors]
```

After seeing how the CRTP helps achieve different goals, let's look at the use of the CRTP in the C++ standard library.

The CRTP in the standard library

The standard library contains a helper type called `std::enabled_shared_from_this` (in the `<memory>` header) that enables objects managed by a `std::shared_ptr` to generate more `std::shared_ptr` instances in a safe manner. The `std::enabled_shared_from_this` class is the base class in the CRTP pattern. However, the previous description may sound abstract, so let's try to understand it with examples.

Let's suppose we have a class called `building` and we are creating `std::shared_ptr` objects in the following manner:

```
struct building {};

building* b = new building();
std::shared_ptr<building> p1{ b }; // [1]
std::shared_ptr<building> p2{ b }; // [2] bad
```

We have a raw pointer and, on line `[1]`, we instantiate a `std::shared_ptr` object to manage its lifetime. However, on line `[2]`, we instantiate a second `std::shared_ptr` object for the same pointer. Unfortunately, the two smart pointers know nothing of each other, so upon getting out of scope, they will both delete the `building` object allocated on the heap. Deleting an object that was already deleted is undefined behavior and will likely result in a crash of the program.

The `std::enable_shared_from_this` class helps us create more `shared_ptr` objects from an existing one in a safe manner. First, we need to implement the CRTP pattern as follows:

```
struct building : std::enable_shared_from_this<building>
{
};
```

Having this new implementation, we can call the member function `shared_from_`
`this` to create more `std::shared_ptr` instances from an object, which all refer to the
same instance of the object:

```
building* b = new building();
std::shared_ptr<building> p1{ b };      // [1]
std::shared_ptr<building> p2{
    b->shared_from_this()};             // [2] OK
```

The interface of the `std::enable_shared_from_this` is as follows:

```
template <typename T>
class enable_shared_from_this
{
public:
  std::shared_ptr<T>        shared_from_this();
  std::shared_ptr<T const>  shared_from_this() const;
  std::weak_ptr<T>          weak_from_this() noexcept;
  std::weak_ptr<T const>    weak_from_this() const noexcept;
  enable_shared_from_this<T>& operator=(
      const enable_shared_from_this<T> &obj ) noexcept;
};
```

The previous example shows how `enable_shared_from_this` works but does not
help understand when it is appropriate to use it. Therefore, let's modify the example to
show a realistic example.

Let's consider that the buildings we have can be upgraded. This is a process that takes
some time and involves several steps. This task, as well as other tasks in the game, are
executed by a designated entity, which we will call `executor`. In its simplest form,
this `executor` class has a public member function called `execute` that takes
a function object and executes it on a different thread. The following listing is
a simple implementation:

```
struct executor
{
    void execute(std::function<void(void)> const& task)
    {
        threads.push_back(std::thread([task]() {
            using namespace std::chrono_literals;
```

```
            std::this_thread::sleep_for(250ms);
            task();
        }));
    }


    ~executor()
    {
        for (auto& t : threads)
            t.join();
    }
private:
    std::vector<std::thread> threads;
};
```

The `building` class has a pointer to an `executor`, which is passed from the client. It also has a member function called `upgrade` that kicks off the execution process. However, the actual upgrade occurs in a different, private, function called `do_upgrade`. This is called from a lambda expression that is passed to the `execute` member function of the `executor`. All these are shown in the following listing:

```
struct building
{
    building()  { std::cout << "building created\n"; }
    ~building() { std::cout << "building destroyed\n"; }


    void upgrade()
    {
        if (exec)
        {
            exec->execute([self = this]() {
                self->do_upgrade();
            });
        }
    }


    void set_executor(executor* e) { exec = e; }
private:
```

```cpp
    void do_upgrade()
    {
        std::cout << "upgrading\n";
        operational = false;

        using namespace std::chrono_literals;
        std::this_thread::sleep_for(1000ms);

        operational = true;
        std::cout << "building is functional\n";
    }

    bool operational = false;
    executor* exec = nullptr;
};
```

The client code is relatively simple: create an executor, create a building managed by a shared_ptr, set the executor reference, and run the upgrade process:

```cpp
int main()
{
    executor e;
    std::shared_ptr<building> b =
        std::make_shared<building>();
    b->set_executor(&e);
    b->upgrade();

    std::cout << "main finished\n";
}
```

If you run this program, you get the following output:

```
building created
main finished
building destroyed
upgrading
building is functional
```

What we can see here is that the building is destroyed before the upgrade process begins. This incurs undefined behavior and, although this program didn't crash, a real-world program would certainly crash.

The culprit for this behavior is this particular line in the upgrading code:

```
exec->execute([self = this]() {
    self->do_upgrade();
});
```

We are creating a lambda expression that captures the `this` pointer. The pointer is later used after the object it points to has been destroyed. To avoid this, we would need to create and capture a `shared_ptr` object. The safe way to do that is with the help of the `std::enable_shared_from_this` class. There are two changes that need to be done. The first is to actually derive the `building` class from the `std::enable_shared_from_this` class:

```
struct building : std::enable_shared_from_this<building>
{
    /* ... */
};
```

The second change requires us to call `shared_from_this` in the lambda capture:

```
exec->execute([self = shared_from_this()]() {
    self->do_upgrade();
});
```

These are two slight changes to our code but the effect is significant. The building object is no longer destroyed before the lambda expression gets executed on a separate thread (because there is now an extra shared pointer that refers to the same object as the shared pointer created in the main function). As a result, we get the output we expected (without any changes to the client code):

```
building created
main finished
upgrading
building is functional
building destroyed
```

You could argue that after the main `function` finishes, we shouldn't care what happens. Mind that this is just a demo program, and in practice, this happens in some other function and the program continues to run long after that function returns.

With this, we conclude the discussion around the curiously recurring template pattern. Next, we will look at a technique called **mixins** that is often mixed with the CRTP pattern.

Mixins

Mixins are small classes that are designed to add functionality to other classes. If you read about mixins, you will often find that the curiously recurring template pattern is used to implement mixins in C++. This is an incorrect statement. The CRTP helps achieve a similar goal to mixins, but they are different techniques. The point of mixins is that they are supposed to add functionality to classes without being a base class to them, which is the key to the CRTP pattern. Instead, mixins are supposed to inherit from the classes they add functionality to, which is the CRTP upside down.

Remember the earlier example with knights and mages that could move forth and back with the `step_forth` and `step_back` member functions? The `knight` and `mage` classes were derived from the `movable_unit` class template that added the functions `advance` and `retreat`, which enabled units to move several steps forth or back. The same example can be implemented using mixins in a reverse order. Here is how:

```
struct knight
{
    void step_forth()
    {
        std::cout << "knight moves forward\n";
    }

    void step_back()
    {
        std::cout << "knight moves back\n";
    }
};

struct mage
{
    void step_forth()
    {
```

```
        std::cout << "mage moves forward\n";
    }

    void step_back()
    {
        std::cout << "mage moves back\n";
    }
};

template <typename T>
struct movable_unit : T
{
    void advance(size_t steps)
    {
        while (steps--)
            T::step_forth();
    }

    void retreat(size_t steps)
    {
        while (steps--)
            T::step_back();
    }
};
```

You will notice that `knight` and `mage` are now classes that don't have any base class. They both provide the `step_forth` and `step_back` member functions just as they did before, when we implemented the CRTP pattern. Now, the `movable_unit` class template is derived from one of these classes and defines the `advance` and `retreat` functions, which call `step_forth` and `step_back` in a loop. We can use them as follows:

```
movable_unit<knight> k;
k.advance(3);
k.retreat(2);

movable_unit<mage> m;
```

```
m.advance(5);
m.retreat(3);
```

This is very similar to what we had with the CRTP pattern, except that now we create instances of movable_unit<knight> and movable_unit<mage> instead of knight and mage. A comparison of the two patterns is shown in the following diagram (with CRTP on the left and mixins on the right):

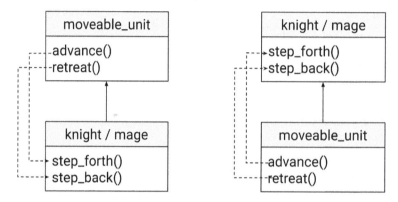

Figure 7.1: Comparison of the CRTP and the mixins patterns

We can combine the static polymorphism achieved with mixins with dynamic polymorphism achieved with interfaces and virtual functions. We'll demonstrate this with the help of an example concerning game units that fight. We had an earlier example when we discussed the CRTP, where the knight and mage classes had a member function called attack.

Let's say we want to define multiple attacking styles. For instance, each game unit can use either an aggressive or a moderate attacking style. So that means four combinations: aggressive and moderate knights, and aggressive and moderate mages. On the other hand, both knights and mages could be lone warriors that are comfortable to fight alone, or are team players that always fight in a group with other units.

That means we could have lone aggressive knights and lone moderate knights as well as team player aggressive knights and team player moderate knights. The same applies to mages. As you can see, the number of combinations grows a lot and mixins are a good way to provide this added functionality without expanding the knight and mage classes. Finally, we want to be able to treat all these polymorphically at runtime. Let's see how we can do this.

First, we can define aggressive and moderate fighting styles. These could be as simple as the following:

```
struct aggressive_style
{
    void fight()
    {
        std::cout << "attack! attack attack!\n";
    }
};

struct moderate_style
{
    void fight()
    {
        std::cout << "attack then defend\n";
    }
};
```

Next, we define mixins as the requirement of being able to fight alone or in a group. These classes are templates and are derived from their template argument:

```
template <typename T>
struct lone_warrior : T
{
    void fight()
    {
        std::cout << "fighting alone.";
        T::fight();
    }
};

template <typename T>
struct team_warrior : T
{
    void fight()
    {
        std::cout << "fighting with a team.";
```

```
            T::fight();
    }
};
```

Last, we need to define the knight and mage classes. These themselves will be mixins for the fighting styles. However, to be able to treat them polymorphically at runtime, we derive them from a base game_unit class that contains a pure virtual method called attack that these classes implement:

```
struct game_unit
{
    virtual void attack() = 0;
    virtual ~game_unit() = default;
};

template <typename T>
struct knight : T, game_unit
{
    void attack()
    {
        std::cout << "draw sword.";
        T::fight();
    }
};

template <typename T>
struct mage : T, game_unit
{
    void attack()
    {
        std::cout << "spell magic curse.";
        T::fight();
    }
};
```

The `knight` and `mage` implementation of the `attack` member function makes use of the `T::fight` method. You have probably noticed that both the `aggresive_style` and `moderate_style` classes on one hand and the `lone_warrior` and `team_warrior` mixin classes on the other hand provide such a member function. This means we can do the following combinations:

```cpp
std::vector<std::unique_ptr<game_unit>> units;

units.emplace_back(new knight<aggressive_style>());
units.emplace_back(new knight<moderate_style>());
units.emplace_back(new mage<aggressive_style>());
units.emplace_back(new mage<moderate_style>());
units.emplace_back(
    new knight<lone_warrior<aggressive_style>>());
units.emplace_back(
    new knight<lone_warrior<moderate_style>>());
units.emplace_back(
    new knight<team_warrior<aggressive_style>>());
units.emplace_back(
    new knight<team_warrior<moderate_style>>());
units.emplace_back(
    new mage<lone_warrior<aggressive_style>>());
units.emplace_back(
    new mage<lone_warrior<moderate_style>>());
units.emplace_back(
    new mage<team_warrior<aggressive_style>>());
units.emplace_back(
    new mage<team_warrior<moderate_style>>());

for (auto& u : units)
    u->attack();
```

In total, there are 12 combinations that we defined here. And this was all possible with only six classes. This shows how mixins help us add functionality while keeping the complexity of the code at a reduced level. If we run the code, we get the following output:

```
draw sword.attack! attack attack!
draw sword.attack then defend
spell magic curse.attack! attack attack!
spell magic curse.attack then defend
draw sword.fighting alone.attack! attack attack!
draw sword.fighting alone.attack then defend
draw sword.fighting with a team.attack! attack attack!
draw sword.fighting with a team.attack then defend
spell magic curse.fighting alone.attack! attack attack!
spell magic curse.fighting alone.attack then defend
spell magic curse.fighting with a team.attack! attack attack!
spell magic curse.fighting with a team.attack then defend
```

We have looked here at two patterns, CRTP and mixins, that are both intended to add additional (common) functionality to other classes. However, although they look similar, they have opposite structures and should not be confused with one another. An alternative technique to leverage common functionalities from unrelated types is called type erasure, which we will discuss next.

Type erasure

The term **type erasure** describes a pattern in which type information is removed, allowing types that are not necessarily related to be treated in a generic way. This is not something specific to the C++ language. This concept exists in other languages with better support than in C++ (such as Python and Java). There are different forms of type erasure such as polymorphism and using `void` pointers (a legacy of the C language, which is to be avoided), but true type erasure is achieved with templates. Before we discuss this, let's briefly look at the others.

The most rudimentary form of type erasure is the use of `void` pointers. This is typical of C and although possible in C++, it is in no way recommended. It is not type-safe and, therefore, error-prone. However, for the sake of the discussion, let's have a look at such an approach.

Let's say we again have `knight` and `mage` types and they both have an attack function (a behavior), and we want to treat them in a common way to exhibit this behavior. Let's see the classes first:

```
struct knight
{
    void attack() { std::cout << "draw sword\n"; }
};

struct mage
{
    void attack() { std::cout << "spell magic curse\n"; }
};
```

In a C-like implementation, we could have a function for each of these types, taking a `void*` to an object of the type, casting it to the expected type of pointer, and then invoking the `attack` member function:

```
void fight_knight(void* k)
{
    reinterpret_cast<knight*>(k)->attack();
}

void fight_mage(void* m)
{
    reinterpret_cast<mage*>(m)->attack();
}
```

These have a similar signature; the only thing that differs is the name. So, we can define a function pointer and then associate an object (or more precisely a pointer to an object) with a pointer to the right function handling it. Here is how:

```
using fight_fn = void(*)(void*);

void fight(
    std::vector<std::pair<void*, fight_fn>> const& units)
{
    for (auto& u : units)
```

```
    {
        u.second(u.first);
    }
}
```

There is no information about types in this last snippet. All that has been erased using
void pointers. The fight function can be invoked as follows:

```
knight k;
mage m;

std::vector<std::pair<void*, fight_fn>> units {
    {&k, &fight_knight},
    {&m, &fight_mage},
};


fight(units);
```

From a C++ perspective, this will probably look odd. It should. In this example, I have
combined C techniques with C++ classes. Hopefully, we will not see snippets of code like
this in production. Things will go wrong by a simple typing error if you pass a mage to
the fight_knight function or the other way around. Nevertheless, it's possible and is
a form of type erasure.

An obvious alternative solution in C++ is using polymorphism through inheritance.
This is the very first solution we saw at the beginning of this chapter. For convenience,
I'll reproduce it here again:

```
struct game_unit
{
    virtual void attack() = 0;
};


struct knight : game_unit
{
    void attack() override
    { std::cout << "draw sword\n"; }
};
```

```
struct mage : game_unit
{
    void attack() override
    { std::cout << "spell magic curse\n"; }
};


void fight(std::vector<game_unit*> const & units)
{
    for (auto unit : units)
        unit->attack();
}
```

The `fight` function can handle `knight` and `mage` objects homogenously. It knows nothing of the actual objects whose addresses were passed to it (within a `vector`). However, it can be argued that types have not been completely erased. Both `knight` and `mage` are `game_unit` and the `fight` function handles anything that is a `game_unit`. For another type to be handled by this function, it needs to derive from the `game_unit` pure abstract class.

And sometimes that's not possible. Perhaps we want to treat unrelated types in a similar matter (a process called **duck typing**) but we are not able to change those types. For instance, we do not own the source code. The solution to this problem is true type erasure with templates.

Before we get to see what this pattern looks like, let's take it step by step to understand how the pattern developed, starting with the unrelated `knight` and `mage`, and the premise that we cannot modify them. However, we can write wrappers around them that would provide a uniform interface to the common functionality (behavior):

```
struct knight
{
    void attack() { std::cout << "draw sword\n"; }
};


struct mage
{
    void attack() { std::cout << "spell magic curse\n"; }
};


struct game_unit
```

```
{
    virtual void attack() = 0;
    virtual ~game_unit() = default;
};

struct knight_unit : game_unit
{
    knight_unit(knight& u) : k(u) {}
    void attack() override { k.attack(); }\

private:
    knight& k;
};

struct mage_unit : game_unit
{
    mage_unit(mage& u) : m(u) {}
    void attack() override { m.attack(); }

private:
    mage& m;
};

void fight(std::vector<game_unit*> const & units)
{
    for (auto u : units)
        u->attack();
}
```

We do not need to call the attack member function in game_unit the same as it was in knight and mage. It can have any name. This choice was purely made on the grounds of mimicking the original behavior name. The fight function takes a collection of pointers to game_unit, therefore being able to handle both knight and mage objects homogenously, as shown next:

```
knight k;
mage m;
```

```
knight_unit ku{ k };
mage_unit mu{ m };

std::vector<game_unit*> v{ &ku, &mu };
fight(v);
```

The trouble with this solution is that there is a lot of duplicate code. The `knight_unit` and `mage_unit` classes are mostly the same. And when other classes need to be handled similarly, this duplication increases more. The solution to code duplication is using templates. We replace `knight_unit` and `mage_unit` with the following class template:

```
template <typename T>
struct game_unit_wrapper : public game_unit
{
    game_unit_wrapper(T& unit) : t(unit) {}

    void attack() override { t.attack(); }
private:
    T& t;
};
```

There is only one copy of this class in our source code but the compiler will instantiate multiple specializations based on its usage. Any type information has been erased, with the exception of some type restrictions—the `T` type must have a member function called `attack` that takes no arguments. Notice that the `fight` function didn't change at all. The client code needs to be slightly changed though:

```
knight k;
mage m;

game_unit_wrapper ku{ k };
game_unit_wrapper mu{ m };

std::vector<game_unit*> v{ &ku, &mu };
fight(v);
```

This leads us to the form of the type erasure pattern by putting the abstract base class and wrapper class template within another class:

```cpp
struct game
{
    struct game_unit
    {
        virtual void attack() = 0;
        virtual ~game_unit() = default;
    };

    template <typename T>
    struct game_unit_wrapper : public game_unit
    {
        game_unit_wrapper(T& unit) : t(unit) {}

        void attack() override { t.attack(); }
    private:
        T& t;
    };

    template <typename T>
    void addUnit(T& unit)
    {
        units.push_back(
            std::make_unique<game_unit_wrapper<T>>(unit));
    }

    void fight()
    {
        for (auto& u : units)
            u->attack();
    }
private:
    std::vector<std::unique_ptr<game_unit>> units;
};
```

The game class contains a collection of game_unit objects and has a method for adding new wrappers to any game unit (that has an attack member function). It also has a member function, fight, to invoke the common behavior. The client code is, this time, the following:

```
knight k;
mage m;

game g;
g.addUnit(k);
g.addUnit(m);

g.fight();
```

In the type erasure pattern, the abstract base class is called a **concept** and the wrapper that inherits from it is called a **model**. If we were to implement the type erasure pattern in the established formal manner it would look as follows:

```
struct unit
{
    template <typename T>
    unit(T&& obj) :
        unit_(std::make_shared<unit_model<T>>(
                std::forward<T>(obj)))
    {}

    void attack()
    {
        unit_->attack();
    }

    struct unit_concept
    {
        virtual void attack() = 0;
        virtual ~unit_concept() = default;
    };

    template <typename T>
```

```
    struct unit_model : public unit_concept
    {
        unit_model(T& unit) : t(unit) {}

        void attack() override { t.attack(); }
    private:
        T& t;
    };

private:
    std::shared_ptr<unit_concept> unit_;
};


void fight(std::vector<unit>& units)
{
    for (auto& u : units)
        u.attack();
}
```

In this snippet, game_unit was renamed as unit_concept and game_unit_ wrapper was renamed as unit_model. There is no other change to them apart from the name. They are members of a new class called unit that stores a pointer to an object that implements unit_concept; that could be unit_model<knight> or unit_ model<mage>. The unit class has a template constructor that enables us to create such model objects from knight and mage objects.

It also has a public member function, attack (again, this can have any name). On the other hand, the fight function handles unit objects and invokes their attack member function. The client code may look as follows:

```
knight k;
mage m;

std::vector<unit> v{ unit(k), unit(m) };

fight(v);
```

If you're wondering where this pattern is used in real-world code, there are two examples in the standard library itself:

- `std::function`: This is a general-purpose polymorphic function wrapper that enables us to store, copy, and invoke anything that is callable, such as functions, lambda expressions, bind expressions, function objects, pointers to member functions, and pointers to data members. Here is an example of using `std::function`:

```cpp
class async_bool
{
    std::function<bool()> check;
public:
    async_bool() = delete;
    async_bool(std::function<bool()> checkIt)
        : check(checkIt)
    { }

    async_bool(bool val)
        : check([val]() {return val; })
    { }

    operator bool() const { return check(); }
};

async_bool b1{ false };
async_bool b2{ true };
async_bool b3{ []() { std::cout << "Y/N? ";
                      char c; std::cin >> c;
                      return c == 'Y' || c == 'y'; } };

if (b1) { std::cout << "b1 is true\n"; }
if (b2) { std::cout << "b2 is true\n"; }
if (b3) { std::cout << "b3 is true\n"; }
```

- `std::any`: This is a class that represents a container to any value of a type that is copy-constructible. An example is used in the following snippet:

```
std::any u;

u = knight{};
if (u.has_value())
    std::any_cast<knight>(u).attack();

u = mage{};
if (u.has_value())
    std::any_cast<mage>(u).attack();
```

Type erasure is an idiom that combines inheritance from object-oriented programming with templates to create wrappers that can store any type. In this section, we have seen how the pattern looks and how it works, as well as some real-world implementations of the pattern.

Next in this chapter, we will discuss a technique called tag dispatching.

Tag dispatching

Tag dispatching is a technique that enables us to select one or another function overload at compile time. It is an alternative to `std::enable_if` and **SFINAE** and is simple to understand and use. The term *tag* describes an empty class that has no members (data), or functions (behavior). Such a class is only used to define a parameter (usually the last) of a function to decide whether to select it at compile-time, depending on the supplied arguments. To better understand this, let's consider an example.

The standard library contains a utility function called `std::advance` that looks as follows:

```
template<typename InputIt, typename Distance>
void advance(InputIt& it, Distance n);
```

Notice that in C++17, this is also `constexpr` (more about this, shortly). This function increments the given iterator by n elements. However, there are several categories of iterators (input, output, forward, bidirectional, and random access). That means such an operation can be computed differently:

- For input iterators, it could call `operator++` a number of n times.

- For bidirectional iterators, it could call either `operator++` a number of n times (if n is a positive number) or `operator--` a number of n times (if n is a negative number).

- For random-access iterators, it can use the `operator+=` to increment it directly with n elements.

This implies there can be three different implementations, but it should be possible to select at compile-time which one is the best match for the category of the iterator it is called for. A solution for this is tag dispatching. And the first thing to do is define the tags. As mentioned earlier, tags are empty classes. Therefore, tags that correspond to the five iterator types can be defined as follows:

```
struct input_iterator_tag {};
struct output_iterator_tag {};
struct forward_iterator_tag : input_iterator_tag {};
struct bidirectional_iterator_tag :
   forward_iterator_tag {};
struct random_access_iterator_tag :
   bidirectional_iterator_tag {};
```

This is exactly how they are defined in the C++ standard library, in the `std` namespace. These tags will be used to define an additional parameter for each overload of `std::advance`, as shown next:

```
namespace std
{
   namespace details
   {
      template <typename Iter, typename Distance>
      void advance(Iter& it, Distance n,
                   std::random_access_iterator_tag)
      {
         it += n;
```

```
        }

        template <typename Iter, typename Distance>
        void advance(Iter& it, Distance n,
                    std::bidirectional_iterator_tag)
        {
            if (n > 0)
            {
                while (n--) ++it;
            }
            else
            {
                while (n++) --it;
            }
        }

        template <typename Iter, typename Distance>
        void advance(Iter& it, Distance n,
                    std::input_iterator_tag)
        {
            while (n--)
            {
                ++it;
            }
        }
    }
}
```

These overloads are defined in a separate (inner) namespace of the std namespace so that the standard namespace is not polluted with unnecessary definitions. You can see here that each of these overloads has three parameters: a reference to an iterator, a number of elements to increment (or decrement), and a tag.

The last thing to do is provide a definition of an `advance` function that is intended for direct use. This function does not have a third parameter but calls one of these overloads by determining the category of the iterator it is called with. Its implementation may look as follows:

```
namespace std
{
    template <typename Iter, typename Distance>
    void advance(Iter& it, Distance n)
    {
        details::advance(it, n,
            typename std::iterator_traits<Iter>::
                            iterator_category{});
    }
}
```

The `std::iterator_traits` class seen here defines a sort of interface for iterator types. For this purpose, it contains several member types, one of them being `iterator_category`. This resolves to one of the iterator tags defined earlier, such as `std::input_iterator_tag` for input iterators or `std::random_access_iterator_tag` for random access iterators. Therefore, based on the category of the supplied iterator, it instantiates one of these tag classes, determining the selection at compile-time of the appropriate overloaded implementation from the `details` namespace. We can invoke the `std::advance` function as follows:

```
std::vector<int> v{ 1,2,3,4,5 };
auto sv = std::begin(v);
std::advance(sv, 2);

std::list<int> l{ 1,2,3,4,5 };
auto sl = std::begin(l);
std::advance(sl, 2);
```

The category type of the `std::vector`'s iterators is random access. On the other hand, the iterator category type for `std::list` is bidirectional. However, we can use a single function that relies on different optimized implementations by leveraging the technique of tag dispatching.

Alternatives to tag dispatching

Prior to C++17, the only alternative to tag dispatching was SFINAE with `enable_if`. We have discussed this topic in *Chapter 5, Type Traits and Conditional Compilation*. This is a rather legacy technique that has better alternatives in modern C++. These alternatives are **constexpr if** and **concepts**. Let's discuss them one at a time.

Using constexpr if

C++11 introduced the concept of `constexpr` values, which are values known at compile-time but also `constexpr` functions that are functions that could be evaluated at compile-time (if all inputs are compile-time values). In C++14, C++17, and C++20, many standard library functions or member functions of standard library classes have been changed to be `constexpr`. One of these is `std::advance`, whose implementation in C++17 is based on the constexpr if feature, also added in C++17 (which was discussed in *Chapter 5, Type Traits and Conditional Compilation*).

The following is a possible implementation in C++17:

```cpp
template<typename It, typename Distance>
constexpr void advance(It& it, Distance n)
{
   using category =
      typename std::iterator_traits<It>::iterator_category;
   static_assert(std::is_base_of_v<std::input_iterator_tag,
                                   category>);

   auto dist =
      typename std::iterator_traits<It>::difference_type(n);
   if constexpr (std::is_base_of_v<
                    std::random_access_iterator_tag,
                    category>)
   {
      it += dist;
   }
   else
   {
      while (dist > 0)
      {
```

```
        --dist;
        ++it;
    }
    if constexpr (std::is_base_of_v<
                        std::bidirectional_iterator_tag,
                        category>)
    {
        while (dist < 0)
        {
            ++dist;
            --it;
        }
    }
    }
}
```

Although this implementation still uses the iterator tags that we saw earlier, they are no longer used to invoke different overloaded functions but to determine the value of some compile-time expressions. The `std::is_base_of` type trait (through the `std::is_base_of_v` variable template) is used to determine the type of the iterator category at compile-time.

This implementation has several advantages:

- Has a single implementation of the algorithm (in the `std` namespace)
- Does not require multiple overloads with implementation details defined in a separate namespace

The client code is unaffected. Therefore, library implementors were able to replace the previous version based on tag dispatching with the new version based on constexpr if, without affecting any line of code calling `std::advance`.

However, in C++20 there is an even better alternative. Let's explore it next.

Using concepts

The previous chapter was dedicated to constraints and concepts, introduced in C++20. We have seen not only how these features work but also some of the concepts that the standard library defines in several headers such as `<concepts>` and `<iterator>`. Some of these concepts specify that a type is some iterator category. For instance, `std::input_iterator` specifies that a type is an input iterator. Similarly, the following concepts are also defined: `std::output_iterator`, `std::forward_iterator`, `std::bidirectional_iterator`, `std::random_access_iterator`, and `std::contiguous_iterator` (the last one indicating that an iterator is a random-access iterator, referring to elements that are stored contiguously in memory).

The `std::input_iterator` concept is defined as follows:

```
template<class I>
    concept input_iterator =
        std::input_or_output_iterator<I> &&
        std::indirectly_readable<I> &&
        requires { typename /*ITER_CONCEPT*/<I>; } &&
        std::derived_from</*ITER_CONCEPT*/<I>,
                            std::input_iterator_tag>;
```

Without getting into too many details, it is worth noting that this concept is a set of constraints that verify the following:

- The iterator is dereferenceable (supports `*i`) and is incrementable (supports `++i` and `i++`).

- The iterator category is derived from `std::input_iterator_tag`.

This means that the category check is performed within the constraint. Therefore, these concepts are still based on the iterator tags, but the technique is significantly different than tag dispatching. As a result, in C++20, we could have yet another implementation for the `std::advance` algorithm, as follows:

```
template <std::random_access_iterator Iter, class Distance>
void advance(Iter& it, Distance n)
{
    it += n;
}


template <std::bidirectional_iterator Iter, class Distance>
```

```cpp
void advance(Iter& it, Distance n)
{
    if (n > 0)
    {
        while (n--) ++it;
    }
    else
    {
        while (n++) --it;
    }
}

template <std::input_iterator Iter, class Distance>
void advance(Iter& it, Distance n)
{
    while (n--)
    {
        ++it;
    }
}
```

There are a couple of things to notice here:

- There are yet again three different overloads of the advanced function.

- These overloads are defined in the std namespace and do not require a separate namespace to hide implementation details.

Although we explicitly wrote several overloads again, this solution is arguably easier to read and understand than the one based on constexpr if because the code is nicely separated into different units (functions), making it easier to follow.

Tag dispatching is an important technique for selecting between overloads at compile-time. It has its trade-offs but also better alternatives if you are using C++17 or C++20. If your compiler supports concepts, you should prefer this alternative for the reasons mentioned earlier.

The next pattern we will look at in this chapter is expression templates.

Expression templates

Expression templates are a metaprogramming technique that enables lazy evaluation of a computation at compile-time. This helps to avoid inefficient operations that occur at runtime. However, this does not come for free, as expression templates require more code and can be cumbersome to read or understand. They are often used in the implementation of linear algebra libraries.

Before seeing how expression templates are implemented, let's understand what is the problem they solve. For this, let's suppose we want to do some operations with matrices, for which we implemented the basic operations, addition, subtraction, and multiplication (either of two matrices or of a scalar and a matrix). We can have the following expressions:

```
auto r1 = m1 + m2;
auto r2 = m1 + m2 + m3;
auto r3 = m1 * m2 + m3 * m4;
auto r4 = m1 + 5 * m2;
```

In this snippet, m1, m2, m3, and m4 are matrices; similarly, r1, r2, r3, and r4 are matrices that result from performing the operations on the right side. The first operation does not pose any problems: m1 and m2 are added and the result is assigned to r1. However, the second operation is different because there are three matrices that are added. That means m1 and m2 are added first and a temporary is created, which is then added to m3 and the result assigned to r2.

For the third operation, there are two temporaries: one for the result of multiplying m1 and m2 and one for the result of multiplying m3 and m4; these two are then added and the result is assigned to r3. Finally, the last operation is similar to the second, meaning that a temporary object results from the multiplication between the scalar 5 and the matrix m2, and then this temporary is added to m1 and the result assigned to r4.

The more complex the operation, the more temporaries are generated. This can affect performance when the objects are large. Expression templates help to avoid this by modeling the computation as a compile-time expression. The entire mathematical expression (such as m1 + 5 * m2) becomes a single expression template computed when the assignment is evaluated and without the need for any temporary object.

To demonstrate this, we will build some examples using vectors not matrices because these are simpler data structures, and the point of the exercise is not to focus on the representation of data but on the creation of expression templates. In the following listing, you can see a minimal implementation of a vector class that provides several operations:

- Constructing an instance from an initializer list or from a value representing a size (no initializing values)
- Retrieving the number of elements in the vector
- Element access with the subscript operator ([ ])

The code goes as follows:

```
template<typename T>
struct vector
{
    vector(std::size_t const n) : data_(n) {}

    vector(std::initializer_list<T>&& l) : data_(l) {}

    std::size_t size() const noexcept
    {
        return data_.size();
    }

    T const & operator[](const std::size_t i) const
    {
        return data_[i];
    }

    T& operator[](const std::size_t i)
    {
        return data_[i];
    }

private:
    std::vector<T> data_;
};
```

This looks very similar to the `std::vector` standard container, and, in fact, it uses this container internally to hold the data. However, this aspect is irrelevant to the problem we want to solve. Remember we are using a vector and not a matrix because it's easier to represent in a few lines of code. Having this class, we can define the necessary operations: addition and multiplication, both between two vectors and between a scalar and a vector:

```cpp
template<typename T, typename U>
auto operator+ (vector<T> const & a, vector<U> const & b)
{
    using result_type = decltype(std::declval<T>() +
                                  std::declval<U>());
    vector<result_type> result(a.size());
    for (std::size_t i = 0; i < a.size(); ++i)
    {
        result[i] = a[i] + b[i];
    }
    return result;
}


template<typename T, typename U>
auto operator* (vector<T> const & a, vector<U> const & b)
{
    using result_type = decltype(std::declval<T>() +
                                  std::declval<U>());
    vector<result_type> result(a.size());
    for (std::size_t i = 0; i < a.size(); ++i)
    {
        result[i] = a[i] * b[i];
    }
    return result;
}


template<typename T, typename S>
auto operator* (S const& s, vector<T> const& v)
{
    using result_type = decltype(std::declval<T>() +
                                  std::declval<S>());
```

```
    vector<result_type> result(v.size());
    for (std::size_t i = 0; i < v.size(); ++i)
    {
        result[i] = s * v[i];
    }
    return result;
}
```

These implementations are relatively straightforward and should not pose a problem to understand at this point. Both the + and * operators take two vectors of potentially different types, such as vector<int> and vector<double>, and return a vector holding elements of a result type. This is determined by the result of adding two values of the template types T and U, using std::declval. This has been discussed in *Chapter 4, Advanced Template Concepts*. A similar implementation is available for multiplying a scalar and a vector. Having these operators available, we can write the following code:

```
vector<int> v1{ 1,2,3 };
vector<int> v2{ 4,5,6 };
double a{ 1.5 };

vector<double> v3 = v1 + a * v2;          // {7.0, 9.5, 12.0}
vector<int>    v4 = v1 * v2 + v1 + v2; // {9, 17, 27}
```

As previously explained, this will create one temporary object while computing v3 and two temporaries while computing v4. These are exemplified in the following diagrams. The first shows the first computation, v3 = v1 + a * v2:

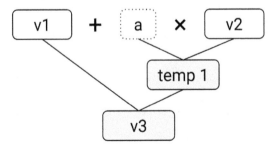

Figure 7.2: A conceptual representation of the first expression

The second diagram, shown next, presents a conceptual representation of the computation of the second expression, v4 = v1 * v2 + v1 + v2:

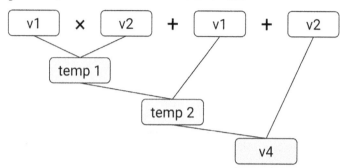

Figure 7.3: A conceptual representation of the second expression

In order to avoid these temporaries, we can rewrite the implementation of the `vector` class using the expression templates pattern. This requires several changes:

- Defining class templates to represent an expression between two objects (such as the expression of adding or multiplying two vectors).

- Modifying the `vector` class and parameterize the container for its internal data, which by default would be a `std::vector` as previously but can also be an expression template.

- Changing the implementation of the overloaded + and * operators.

Let's see how this is done, starting with the vector implementation. Here is the code:

```
template<typename T, typename C = std::vector<T>>
struct vector
{
    vector() = default;

    vector(std::size_t const n) : data_(n) {}

    vector(std::initializer_list<T>&& l) : data_(l) {}

    vector(C const & other) : data_(other) {}

    template<typename U, typename X>
    vector(vector<U, X> const& other) : data_(other.size())
    {
```

```cpp
        for (std::size_t i = 0; i < other.size(); ++i)
            data_[i] = static_cast<T>(other[i]);
    }

    template<typename U, typename X>
    vector& operator=(vector<U, X> const & other)
    {
        data_.resize(other.size());
        for (std::size_t i = 0; i < other.size(); ++i)
            data_[i] = static_cast<T>(other[i]);

        return *this;
    }

    std::size_t size() const noexcept
    {
        return data_.size();
    }

    T operator[](const std::size_t i) const
    {
        return data_[i];
    }

    T& operator[](const std::size_t i)
    {
        return data_[i];
    }

    C& data() noexcept { return data_; }

    C const & data() const noexcept { return data_; }

private:
    C data_;
};
```

In addition to the operations available in the initial implementation, this time we have also defined the following:

- A default constructor

- A conversion constructor from a container

- A copy constructor from a `vector` containing elements of a potentially different type

- A copy-assignment operator from a `vector` containing elements of a potentially different type

- Member function `data` that provides access to the underlaying container holding the data

An expression template is a simple class template that stores two operands and provides a way to perform the evaluation of the operation. In our case, we need to implement expressions for adding two vectors, multiplying two vectors, and multiplying a scalar and a vector. Let's look at the implementation of the expression template for adding two vectors:

```
template<typename L, typename R>
struct vector_add
{
    vector_add(L const & a, R const & b) : lhv(a), rhv(b) {}

    auto operator[](std::size_t const i) const
    {
        return lhv[i] + rhv[i];
    }

    std::size_t size() const noexcept
    {
        return lhv.size();
    }

private:
    L const & lhv;
    R const & rhv;
};
```

This class stores constant references to two vectors (or, in fact, any type that overloads the subscript operator and provides a `size` member function). The evaluation of the expression occurs in the overloaded subscript operator but not for the entire vector; only the elements at the indicated index are added.

Notice that this implementation does not handle vectors of different sizes (which you can take as an exercise to change). However, it should be easy to understand the lazy nature of this approach since the addition operation only occurs when invoking the subscript operator.

The multiplication expression templates for the two operations we need are implemented in a similar fashion. The code is shown in the next listing:

```
template<typename L, typename R>
struct vector_mul
{
    vector_mul(L const& a, R const& b) : lhv(a), rhv(b) {}

    auto operator[](std::size_t const i) const
    {
        return lhv[i] * rhv[i];
    }

    std::size_t size() const noexcept
    {
        return lhv.size();
    }

private:
    L const & lhv;
    R const & rhv;
};

template<typename S, typename R>
struct vector_scalar_mul
{
    vector_scalar_mul(S const& s, R const& b) :
        scalar(s), rhv(b)
```

```
    {}

    auto operator[](std::size_t const i) const
    {
        return scalar * rhv[i];
    }

    std::size_t size() const noexcept
    {
        return rhv.size();
    }

private:
    S const & scalar;
    R const & rhv;
};
```

The last part of the change is to modify the definition of the overloaded + and * operators, which is as follows:

```
template<typename T, typename L, typename U, typename R>
auto operator+(vector<T, L> const & a,
               vector<U, R> const & b)
{
    using result_type = decltype(std::declval<T>() +
                                 std::declval<U>());
    return vector<result_type, vector_add<L, R>>(
        vector_add<L, R>(a.data(), b.data()));
}

template<typename T, typename L, typename U, typename R>
auto operator*(vector<T, L> const & a,
               vector<U, R> const & b)
{
    using result_type = decltype(std::declval<T>() +
                                 std::declval<U>());
    return vector<result_type, vector_mul<L, R>>(
```

```
        vector_mul<L, R>(a.data(), b.data())));
}


template<typename T, typename S, typename E>
auto operator*(S const& a, vector<T, E> const& v)
{
    using result_type = decltype(std::declval<T>() +
                                  std::declval<S>());
    return vector<result_type, vector_scalar_mul<S, E>>(
        vector_scalar_mul<S, E>(a, v.data()));
}
```

Although the code is more complex when implementing this pattern, the client code does not need to change. The snippet showed earlier works without any modifications but in a lazy manner. The evaluation of each element in the result is triggered by the invocation of the subscript operator that occurs in the copy-constructor and copy-assignment operator of the vector class.

If this pattern looks cumbersome to you, there is the alternative of something better: the ranges library.

Using ranges as an alternative to expression templates

One of the major features of C++20 is the ranges library. A *range* is a generalization of a container - a class that allows you to iterate over its data (elements). A key element of the range library is the views. These are non-owning wrappers of other ranges that transform the underlying range through some operation.

Moreover, they are lazy-evaluated and the time to construct, copy, or destroy them does not depend on the size of the underlying range. The lazy evaluation (the fact that transformations are applied to elements at the moment they are requested, not when the view is created) is a key feature of the library. However, that is exactly what the expression templates are also providing. As a result, many uses of the expression templates can be replaced with ranges. Ranges will be discussed in detail in the next chapter.

The C++ ranges library is based on the **range-v3** library created by Eric Niebler. This library is available at https://github.com/ericniebler/range-v3/. Using range-v3, we can write the following code to perform the operation v1 + a * v2:

```
namespace rv = ranges::views;
std::vector<int> v1{ 1, 2, 3 };
```

```
std::vector<int> v2{ 4, 5, 6 };
double a { 1.5 };

auto sv2 = v2 |
            rv::transform([&a](int val) {return a * val; });
auto v3 = rv::zip_with(std::plus<>{}, v1, sv2);
```

There is no need for a custom implementation of a vector class; it just works with the std::vector container. There is no need to overload any operator also. The code should be easy to follow, at least if you have some familiarity with the ranges library. First, we create a view that transforms the elements of the v2 vector by multiplying each element with a scalar. Then, a second view is created that applies the plus operator on the elements of the v1 range and the view resulting from the previous operation.

Unfortunately, this code cannot be written in C++20 using the standard library, because the zip_with view has not been included in C++20. However, this view will be available in C++23 under the name zip_view. Therefore, in C++23, we will be able to write this code as follows:

```
namespace rv = std::ranges::views;
std::vector<int> v1{ 1, 2, 3 };
std::vector<int> v2{ 4, 5, 6 };
double a { 1.5 };

auto sv2 = v2 |
            rv::transform([&a](int val) {return a * val; });
auto v3 = rv::zip_wiew(std::plus<>{}, v1, sv2);
```

To conclude the discussion of the expression templates pattern, you should keep in mind the following takeaways: the pattern is designed to provide lazy evaluation for costly operations, and it does so at the expense of having to write more code (that is also arguably more cumbersome) and increased compile-times (since heavy template code will have an impact on that). However, as of C++20, a good alternative to this pattern is represented by the ranges library. We will learn about this new library in *Chapter 9, The Ranges Library*.

For the next and last section of this chapter, we will look at type lists.

Typelists

A **type list** (also spelled *typelist*) is a compile-time construct that enables us to manage a sequence of types. A typelist is somehow similar to a tuple but does not store any data. A typelist only carries type information and is used exclusively at compile-time for implementing different metaprogramming algorithms, type switches, or design patterns such as *Abstract Factory* or *Visitor*.

> **Important Note**
>
> Although both the *type list* and *typelist* spellings are in use, most of the time you will find the term *typelist* in C++ books and articles. Therefore, this will be the form we will use in this book.

Typelists were popularized by Andrei Alexandrescu in his book, *Modern C++ Design*, published a decade before the release of C++11 (and variadic templates). Alexandrescu defined a typelist as follows:

```
template <class T, class U>
struct Typelist
{
   typedef T Head;
   typedef U Tail;
};
```

In his implementation, a typelist is composed of a head—which is a type, and a tail—which is another typelist. In order to perform various operations on the typelist (which will be discussed shortly) we also need a type to represent the end of the typelist. This can be a simple, empty type that Alexandrescu defined as follows:

```
class null_typelist {};
```

Having these two constructs, we can define typelists in the following way:

```
typedef Typelist<int,
                 Typelist<double, null_typelist>> MyList;
```

Variadic templates make the implementation of typelists simpler, as shown in the next snippet:

```
template <typename ... Ts>
struct typelist {};


using MyList = typelist<int, double>;
```

The implementation of operations of typelists (such as accessing a type at a given index, adding or removing types from the list, and so on) differs significantly depending on the selected approach. In this book, we will only consider the variadic template version. The advantage of this approach is simplicity at different levels: the definition of the typelist is shorter, there is no need for a type to represent the end of the list, and defining typelist aliases is also shorter and easier to read.

Today, perhaps many of the problems for which typelists represented the solution can be also solved using variadic templates. However, there are still scenarios where typelists are required. Here is an example: let's consider a variadic metafunction (a type trait that performs a transformation of types) that does some transformation (such as adding the const qualifier) to the type template arguments. This metafunction defines a member type that represents the input types and one that represents the transformed types. If you try to define it as follows, it will not work:

```
template <typename ... Ts>
struct transformer
{
    using input_types  = Ts...;
    using output_types = std::add_const_t<Ts>...;
};
```

This code produces compiler errors, because the expansion of the parameter pack is not possible in this context. This is a topic we discussed in *Chapter 3, Variadic Templates*. The solution to this is to use a typelist, as follows:

```
template <typename ... Ts>
struct transformer
{
    using input_types  = typelist<Ts...>;
    using output_types = typelist<std::add_const_t<Ts>...>;
};
```

```
static_assert(
    std::is_same_v<
        transformer<int, double>::output_types,
        typelist<int const, double const>>);
```

The change is minimal but produces the expected result. Although this is a good example of where typelists are needed, it's not a typical example of where typelists are used. We will look at such an example next.

Using typelists

It's worth exploring a more complex example before we look at how to implement operations on typelists. This should give you an understanding of the possible usage of typelists, although you can always search for more online.

Let's return to the example of the game units. For simplicity, we'll only consider the following class:

```
struct game_unit
{
    int attack;
    int defense;
};
```

A game unit has two data members representing indices (or levels) for attacking and defending. We want to operate changes on these members with the help of some functors. Two such functions are shown in the following listing:

```
struct upgrade_defense
{
    void operator()(game_unit& u)
    {
        u.defense = static_cast<int>(u.defense * 1.2);
    }
};

struct upgrade_attack
{
    void operator()(game_unit& u)
    {
```

```
        u.attack += 2;
    }
};
```

The first increases the defense index by 20%, while the second increases the attack index by two units. Although this is a small example meant to demonstrate the use case, you can imagine a larger variety of functors like this that could be applied in some well-defined combinations. In our example, however, we want to apply these two functors on a game_unit object. We'd like to have a function as follows:

```
void upgrade_unit(game_unit& unit)
{
    using upgrade_types =
        typelist<upgrade_defense, upgrade_attack>;
    apply_functors<upgrade_types>{}(unit);
}
```

This upgrade_unit function takes a game_unit object and applies the upgrade_defense and upgrade_attack functors to it. For this, it uses another helper functor called apply_functors. This is a class template that has a single template argument. This template argument is a typelist. A possible implementation for the apply_functors functor is shown next:

```
template <typename TL>
struct apply_functors
{
private:
    template <size_t I>
    static void apply(game_unit& unit)
    {
        using F = at_t<I, TL>;
        std::invoke(F{}, unit);
    }

    template <size_t... I>
    static void apply_all(game_unit& unit,
```

```
                         std::index_sequence<I...>)
    {
        (apply<I>(unit), ...);
    }

public:
    void operator()(game_unit& unit) const
    {
        apply_all(unit,
                  std::make_index_sequence<length_v<TL>>{});
    }
};
```

This class template has an overloaded call operator and two private helper functions:

- apply, which applies the functor from the I index of the typelist to a game_unit object.

- apply_all, which applies all the functors in the typelist to a game_unit object by using the apply function in a pack expansion.

We can use the upgrade_unit function as follows:

```
game_unit u{ 100, 50 };
std::cout << std::format("{},{}\n", u.attack, u.defense);
// prints 100,50


upgrade_unit(u);
std::cout << std::format("{},{}\n", u.attack, u.defense);
// prints 102,60
```

If you paid attention to the implementation of the apply_functors class template, you will have noticed the use of the at_t alias template and the length_v variable template, which we have not defined yet. We will look at these two and more in the next section.

Implementing operations on typelists

A typelist is a type that only carries valuable information at compile-time. A typelist acts as a container for other types. When you work with typelists, you need to perform various operations, such as counting the types in the list, accessing a type at a given index, adding a type at the beginning or the end of the list, or the reverse operation, removing a type from the beginning or the end of the list, and so on. If you think about it, these are typical operations you'd use with a container such as a vector. Therefore, in this section, we'll discuss how to implement the following operations:

- `size`: Determines the size of the list
- `front`: Retrieves the first type in the list
- `back`: Retrieves the last type in the list
- `at`: Retrieves the type at the specified index in the list
- `push_back`: Adds a new type to the end of the list
- `push_front`: Adds a new type to the beginning of the list
- `pop_back`: Removes the type at the end of the list
- `pop_front`: Removes the type at the beginning of the list

A typelist is a compile-time construct. It is an immutable entity. Therefore, the operations that add or remove a type do not modify a typelist but create a new one. We'll see that shortly. But first, let's start with the simplest operation, which is retrieving the size of a typelist.

To avoid naming confusion with the `size_t` type, we'll call this operation `lenght_t`, and not `size_t`. We can define this as follows:

```
namespace detail
{
    template <typename TL>
    struct length;

    template <template <typename...> typename TL,
              typename... Ts>
    struct length<TL<Ts...>>
    {
        using type =
            std::integral_constant<std::size_t, sizeof...(Ts)>;
```

```
    };
}

template <typename TL>
using length_t = typename detail::length<TL>::type;

template <typename TL>
constexpr std::size_t length_v = length_t<TL>::value;
```

In the `detail` namespace, we have a class template called `length`. There is a primary template (without a definition) and a specialization for a typelist. This specialization defines a member type called `type` that is a `std::integral_constant`, with a value of the type `std::size_t` representing the number of arguments in the parameter pack `Ts`. Furthermore, we have an alias template, `length_h`, that is an alias for the member called `type` of the `length` class template. Finally, we have a variable template called `length_v` that is initialized from the value of the `std::integral_constant` member, which is also called `value`.

We can verify the correctness of this implementation with the help of some `static_assert` statements, as follows:

```
static_assert(
    length_t<typelist<int, double, char>>::value == 3);
static_assert(length_v<typelist<int, double, char>> == 3);
static_assert(length_v<typelist<int, double>> == 2);
static_assert(length_v<typelist<int>> == 1);
```

The approach used here will be used for defining all the other operations. Let's look next at accessing the front type in the list. This is shown in the next listing:

```
struct empty_type {};

namespace detail
{
    template <typename TL>
    struct front_type;

    template <template <typename...> typename TL,
              typename T, typename... Ts>
    struct front_type<TL<T, Ts...>>
```

```
    {
        using type = T;
    };


    template <template <typename...> typename TL>
    struct front_type<TL<>>
    {
        using type = empty_type;
    };
}


template <typename TL>
using front_t = typename detail::front_type<TL>::type;
```

In the `detail` namespace, we have a class template called `front_type`. Again, we declared a primary template but without a definition. However, we have two specializations: one for a typelist that contains at least one type and one for an empty typelist. In the former case, the `type` member is aliasing the first type in the typelist. In the latter case, there is no type so the `type` member is aliasing a type called `empty_type`. This is an empty class whose only role is to act as the return type for operations where no type is to be returned. We can verify the implementation as follows:

```
static_assert(
    std::is_same_v<front_t<typelist<>>, empty_type>);
static_assert(
    std::is_same_v<front_t<typelist<int>>, int>);
static_assert(
    std::is_same_v<front_t<typelist<int, double, char>>,
                    int>);
```

If you expect the implementation of the operation for accessing the back type to be similar, you will not be disappointed. Here is how it looks:

```
namespace detail
{
    template <typename TL>
    struct back_type;

    template <template <typename...> typename TL,
```

```
            typename T, typename... Ts>
    struct back_type<TL<T, Ts...>>
    {
        using type = back_type<TL<Ts...>>::type;
    };


    template <template <typename...> typename TL,
              typename T>
    struct back_type<TL<T>>
    {
        using type = T;
    };


    template <template <typename...> typename TL>
    struct back_type<TL<>>
    {
        using type = empty_type;
    };
}


template <typename TL>
using back_t = typename detail::back_type<TL>::type;
```

The only significant difference with this implementation is that there are three specializations of the back_type class template and there is recursion involved. The three specializations are for an empty typelist, a typelist with a single type, and a typelist with two or more types. The last one (which is actually the first in the previous listing) is using template recursion in the definition of its type member. We have seen how this works in *Chapter 4, Advanced Template Concepts*. To ensure we implemented the operation the right way we can do some validation as follows:

```
static_assert(
    std::is_same_v<back_t<typelist<>>, empty_type>);
static_assert(
    std::is_same_v<back_t<typelist<int>>, int>);
static_assert(
    std::is_same_v<back_t<typelist<int, double, char>>,
                   char>);
```

Apart from accessing the first and last type in a typelist, we are also interested in accessing a type at any given index. However, the implementation of this operation is less trivial. Let's see it first:

```
namespace detail
{
    template <std::size_t I, std::size_t N, typename TL>
    struct at_type;

    template <std::size_t I, std::size_t N,
              template <typename...> typename TL,
              typename T, typename... Ts>
    struct at_type<I, N, TL<T, Ts...>>
    {
        using type =
            std::conditional_t<
                I == N,
                T,
                typename at_type<I, N + 1, TL<Ts...>>::type>;
    };

    template <std::size_t I, std::size_t N>
    struct at_type<I, N, typelist<>>
    {
        using type = empty_type;
    };
}

template <std::size_t I, typename TL>
using at_t = typename detail::at_type<I, 0, TL>::type;
```

The at_t alias template has two template arguments: an index and a typelist. The at_t template is an alias for the member type of the at_type class template from the detail namespace. The primary template has three template parameters: an index representing the position of the type to retrieve (I), another index representing the current position in the iteration of the types in the list (N), and a typelist (TL).

There are two specializations of this primary template: one for a typelist that contains at least one type and one for an empty typelist. In the latter case, the member `type` is aliasing the `empty_type` type. In the former case, the member `type` is defined with the help of the `std::conditional_t` metafunction. This defines its member `type` as the first type (`T`) when `I == N`, or as the second type (`typename at_type<I, N + 1, TL<Ts...>>::type`) when this condition is false. Here, again, we employ template recursion, incrementing the value of the second index with each iteration. The following `static_assert` statements validate the implementation:

```
static_assert(
    std::is_same_v<at_t<0, typelist<>>, empty_type>);
static_assert(
    std::is_same_v<at_t<0, typelist<int>>, int>);
static_assert(
    std::is_same_v<at_t<0, typelist<int, char>>, int>);

static_assert(
    std::is_same_v<at_t<1, typelist<>>, empty_type>);
static_assert(
    std::is_same_v<at_t<1, typelist<int>>, empty_type>);
static_assert(
    std::is_same_v<at_t<1, typelist<int, char>>, char>);

static_assert(
    std::is_same_v<at_t<2, typelist<>>, empty_type>);
static_assert(
    std::is_same_v<at_t<2, typelist<int>>, empty_type>);
static_assert(
    std::is_same_v<at_t<2, typelist<int, char>>,
                   empty_type>);
```

The next category of operations to implement is adding a type to the beginning and the end of a typelist. We call these `push_back_t` and `push_front_t` and their definitions are as follows:

```
namespace detail
{
    template <typename TL, typename T>
```

```
    struct push_back_type;

    template <template <typename...> typename TL,
              typename T, typename... Ts>
    struct push_back_type<TL<Ts...>, T>
    {
        using type = TL<Ts..., T>;
    };

    template <typename TL, typename T>
    struct push_front_type;

    template <template <typename...> typename TL,
              typename T, typename... Ts>
    struct push_front_type<TL<Ts...>, T>
    {
        using type = TL<T, Ts...>;
    };
}

template <typename TL, typename T>
using push_back_t =
    typename detail::push_back_type<TL, T>::type;

template <typename TL, typename T>
using push_front_t =
    typename detail::push_front_type<TL, T>::type;
```

Based on what we have seen so far with the previous operations, these should be straightforward to understand. The opposite operations, when we remove the first or last type from a typelist, are more complex though. The first one, pop_front_t, looks as follows:

```
namespace detail
{
    template <typename TL>
    struct pop_front_type;
```

```
        template <template <typename...> typename TL,
                  typename T, typename... Ts>
        struct pop_front_type<TL<T, Ts...>>
        {
            using type = TL<Ts...>;
        };

        template <template <typename...> typename TL>
        struct pop_front_type<TL<>>
        {
            using type = TL<>;
        };
    }

    template <typename TL>
    using pop_front_t =
        typename detail::pop_front_type<TL>::type;
```

We have the primary template, `pop_front_type`, and two specializations: the first for a typelist with at least one type, and the second for an empty typelist. The latter defines the member `type` as an empty list; the former defines the member `type` as a typelist with the tail composed from the typelist argument.

The last operation, removing the last type in a typelist, called `pop_back_t`, is implemented as follows:

```
    namespace detail
    {
        template <std::ptrdiff_t N, typename R, typename TL>
        struct pop_back_type;

        template <std::ptrdiff_t N, typename... Ts,
                  typename U, typename... Us>
        struct pop_back_type<N, typelist<Ts...>,
                                typelist<U, Us...>>
        {
            using type =
                typename pop_back_type<N - 1,
```

```
                                      typelist<Ts..., U>,
                                      typelist<Us...>>::type;
    };


    template <typename... Ts, typename... Us>
    struct pop_back_type<0, typelist<Ts...>,
                            typelist<Us...>>
    {
        using type = typelist<Ts...>;
    };


    template <typename... Ts, typename U, typename... Us>
    struct pop_back_type<0, typelist<Ts...>,
                            typelist<U, Us...>>
    {
        using type = typelist<Ts...>;
    };


    template <>
    struct pop_back_type<-1, typelist<>, typelist<>>
    {
        using type = typelist<>;
    };
}


template <typename TL>
using pop_back_t = typename detail::pop_back_type<
    static_cast<std::ptrdiff_t>(length_v<TL>)-1,
                typelist<>, TL>::type;
```

For implementing this operation, we need to start with a typelist and recursively construct another typelist, element by element, until we get to the last type in the input typelist, which should be omitted. For this, we use a counter that tells us how many times we iterated the typelist.

This is initiated with the size of the typelist minus one and we need to stop when we reach zero. For this reason, the pop_back_type class template has four specializations, one for the general case when we are at some iteration in the typelist, two for the case when the counter reached zero, and one for the case when the counter reached the value minus one. This is the case when the initial typelist was empty (therefore, `length_t<TL> - 1` would evaluate to -1). Here are some asserts that show how to use `pop_back_t` and validate its correctness:

```
static_assert(std::is_same_v<pop_back_t<typelist<>>,
                             typelist<>>);
static_assert(std::is_same_v<pop_back_t<typelist<double>>,
                             typelist<>>);
static_assert(
   std::is_same_v<pop_back_t<typelist<double, char>>,
                             typelist<double>>);
static_assert(
   std::is_same_v<pop_back_t<typelist<double, char, int>>,
                             typelist<double, char>>);
```

With these defined, we have provided a series of operations that are necessary for working with typelists. The `length_t` and `at_t` operations were used in the example shown earlier with the execution of functors on `game_unit` objects. Hopefully, this section provided a helpful introduction to typelists and enabled you to understand not only how they are implemented but also how they can be used.

Summary

This chapter was dedicated to learning various metaprogramming techniques. We started by understanding the differences between dynamic and static polymorphism and then looked at the curiously recurring template pattern for implementing the latter.

Mixins was another pattern that has a similar purpose as CRTP—adding functionality to classes, but unlike CRTP, without modifying them. The third technique we learned about was type erasure, which allows similar types that are unrelated to be treated generically. In the second part, we learned about tag dispatching – which allow us to select between overloads at compile time, and expression templates – which enable lazy evaluation of a computation at compile-time to avoid inefficient operations that occur at runtime. Lastly, we explored typelists and learned how they are used and how we can implement operations with them.

In the next chapter, we will look at the core pillars of the standard template library, containers, iterators, and algorithms.

Questions

1. What are the typical problems that can be solved by CRTP?

2. What are mixins and what is their purpose?

3. What is type erasure?

4. What is tag dispatching and what are its alternatives?

5. What are expression templates and where are they used?

Further reading

- *Design Patterns: Elements of Reusable Object-Oriented Software* – Erich Gamma, Richard Helm, Ralph Johnson, John Vlissides, p. 163, Addison-Wesley

- *Modern C++ Design: Generic Programming and Design Patterns Applied* – Andrei Alexandrescu, Addison-Wesley Professional

- *Mixin-Based Programming in C++* – Yannis Smaragdakis, Don Batory, `https://yanniss.github.io/practical-fmtd.pdf`

- *Curiously Recurring Template Patterns* – James Coplien, `http://sites.google.com/a/gertrudandcope.com/info/Publications/InheritedTemplate.pdf`

- *Mixin Classes: The Yang of the CRTP* – Jonathan Boccara, `https://www.fluentcpp.com/2017/12/12/mixin-classes-yang-crtp/`

- *What the Curiously Recurring Template Pattern can bring to your code* – Jonathan Boccara, `https://www.fluentcpp.com/2017/05/16/what-the-crtp-brings-to-code/`

- *Combining Static and Dynamic Polymorphism with C++ Mixin classes* – Michael Afanasiev, `https://michael-afanasiev.github.io/2016/08/03/Combining-Static-and-Dynamic-Polymorphism-with-C++-Template-Mixins.html`

- *Why C++ is not just an Object-Oriented Programming Language* – Bjarne Stroustrup, `https://www.stroustrup.com/oopsla.pdf`

- `enable_shared_from_this` - *overview, examples, and internals* – Hitesh Kumar, `https://www.nextptr.com/tutorial/ta1414193955/enable_shared_from_this-overview-examples-and-internals`

- *Tag dispatch versus concept overloading* – Arthur O'Dwyer, `https://quuxplusone.github.io/blog/2021/06/07/tag-dispatch-and-concept-overloading/`

- *C++ Expression Templates: An Introduction to the Principles of Expression Templates* – Klaus Kreft and Angelika Langer, `http://www.angelikalanger.com/Articles/Cuj/ExpressionTemplates/ExpressionTemplates.htm`

- *We don't need no stinking expression templates*, `https://gieseanw.wordpress.com/2019/10/20/we-dont-need-no-stinking-expression-templates/`

- *Generic Programming: Typelists and Applications* – Andrei Alexandrescu, `https://www.drdobbs.com/generic-programmingtypelists-and-applica/184403813`

- *Of type lists and type switches* – Bastian Rieck, `https://bastian.rieck.me/blog/posts/2015/type_lists_and_switches/`

8
Ranges and Algorithms

By reaching this point of the book, you have learned everything about the syntax and the mechanism of templates in C++, up to the latest version of the standard, C++20. This has equipped you with the necessary knowledge to write templates from simple forms to complex ones. Templates are the key to writing generic libraries. Even though you might not write such a library yourself, you'd still be using one or more. In fact, the everyday code that you're writing in C++ uses templates. And the main reason for that is that as a modern C++ developer, you're using the standard library, which is a library based on templates.

However, the standard library is a collection of many libraries, such as the containers library, iterators library, algorithms library, numeric library, input/output library, filesystem library, regular expressions library, thread support library, utility libraries, and others. Overall, it's a large library that could make the topic of at least an entire book. However, it is worth exploring some key parts of the library to help you get a better understanding of some of the concepts and types you are or could be using regularly.

Because addressing this topic in a single chapter would lead to a significantly large chapter, we will split the discussion into two parts. In this chapter, we will address the following topics:

- Understanding the design of containers, iterators, and algorithms
- Creating a custom container and iterator
- Writing a custom general-purpose algorithm

By the end of this chapter, you will have a good understanding of the three main pillars of the standard template library, which are containers, iterators, and algorithms.

We will begin this chapter with an overview of what the standard library has to offer in this respect.

Understanding the design of containers, iterators, and algorithms

Containers are types that represent collections of elements. These collections can be implemented based on a variety of data structures, each with different semantics: lists, queues, trees, and so on. The standard library provides three categories of containers:

- **Sequence containers**: `vector`, `deque`, `list`, `array`, and `forward_list`
- **Associative containers**: `set`, `map`, `multiset`, and `multimap`
- **Unordered associative containers**: `unordered_set`, `unordered_map`, `unordered_multiset`, and `unordered_multimap`

In addition to this, there are also container adaptors that provide a different interface for sequence containers. This category includes the `stack`, `queue`, and `priority_queue` classes. Finally, there is a class called `span` that represents a non-owning view over a contiguous sequence of objects.

The rationale for these containers to be templates was presented in *Chapter 1, Introduction to Templates*. You don't want to write the same implementation again and again for each different type of element that you need to store in a container. Arguably, the most used containers from the standard library are the following:

- `vector`: This is a variable-size collection of elements stored contiguously in memory. It's the default container you would choose, provided no special requirements are defined. The internal storage expands or shrinks automatically as needed to accommodate the stored elements. The vector allocates more memory than needed so that the risk of having to expand is low. The expansion is a costly operation because new memory needs to be allocated, the content of the current storage needs to be copied to the new one, and lastly, the previous storage needs to be discarded. Because elements are stored contiguously in memory, they can be randomly accessed by an index in constant time.

- `array`: This is a fixed-size collection of elements stored contiguously in memory. The size must be a compile-time constant expression. The semantics of the `array` class is the same as a structure holding a C-style array (`T [n]`). Just like the `vector` type, the elements of the `array` class can be accessed randomly in constant time.

- `map`: This is a collection that associates a value to a unique key. Keys are sorted with a comparison function and the `map` class is typically implemented as a red-black tree. The operations to search, insert, or remove elements have logarithmic complexity.

- `set`: This is a collection of unique keys. The keys are the actual values stored in the container; there are no key-value pairs as in the case of the `map` class. However, just like in the case of the `map` class, `set` is typically implemented as a red-black tree that has logarithmic complexity for searching, inserting, and the removing of elements.

Regardless of their type, the standard containers have a few things in common:

- Several common member types
- An allocator for storage management (with the exception of the `std::array` class)
- Several common member functions (some of them are missing from one or another container)
- Access to the stored data with the help of iterators

The following member types are defined by all standard containers:

```
using value_type       = /* ... */;
using size_type        = std::size_t;
using difference_type  = std::ptrdiff_t;
using reference        = value_type&;
using const_reference  = value_type const&;
using pointer          = /* ... */;
using const_pointer    = /* ... */;
using iterator         = /* ... */;
using const_iterator   = /* ... */;
```

The actual types these names are aliasing may differ from container to container. For instance, for `std::vector`, `value_type` is the template argument `T`, but for `std::map`, `value_type` is the `std::pair<const Key, T>` type. The purpose of these member types is to help with generic programming.

Except for the `std::array` class, which represents an array of a size known at compile time, all the other containers allocate memory dynamically. This is controlled with the help of an object called an **allocator**. Its type is specified as a type template parameter, but all containers default it to `std::allocator` if none is specified. This standard allocator uses the global `new` and `delete` operators for allocating and releasing memory. All constructors of standard containers (including copy and move constructors) have overloads that allow us to specify an allocator.

There are also common member functions defined in the standard containers. Here are some examples:

- `size`, which returns the number of elements (not present in `std::forward_list`).
- `empty`, which checks whether the container is empty.
- `clear`, which clears the content of the container (not present in `std::array`, `std::stack`, `std::queue`, and `std::priority_queue`).
- `swap`, which swaps the content of the container objects.
- `begin` and `end` methods, which return iterators to the beginning and end of the container (not present in `std::stack`, `std::queue`, and `std::priority_queue`, although these are not containers but container adaptors).

The last bullet mentions iterators. These are types that abstract the details of accessing elements in a container, providing a uniform way to identify and traverse the elements of containers. This is important because a key part of the standard library is represented by general-purpose algorithms. There are over one hundred such algorithms, ranging from sequence operations (such as `count`, `count_if`, `find`, and `for_each`) to modifying operations (such as `copy`, `fill`, `transform`, `rotate`, and `reverse`) to partitioning and sorting (`partition`, `sort`, `nth_element`, and more) and others. Iterators are key for ensuring they work generically. If each container had different ways to access its elements, writing generic algorithms would be virtually impossible.

Let's consider the simple operation of copying elements from one container to another. For instance, we have a `std::vector` object and we want to copy its elements to a `std::list` object. This could look as follows:

```
std::vector<int> v {1, 2, 3};
std::list<int> l;


for (std::size_t i = 0; i < v.size(); ++i)
    l.push_back(v[i]);
```

What if we want to copy from a `std::list` to a `std::set`, or from a `std::set` to a `std::array`? Each case would require different kinds of code. However, general-purpose algorithms enable us to do this in a uniform way. The following snippet shows such an example:

```
std::vector<int> v{ 1, 2, 3 };


// copy vector to vector
std::vector<int> vc(v.size());
std::copy(v.begin(), v.end(), vc.begin());


// copy vector to list
std::list<int> l;
std::copy(v.begin(), v.end(), std::back_inserter(l));


// copy list to set
std::set<int> s;
std::copy(l.begin(), l.end(), std::inserter(s, s.begin()));
```

Here we have a std::vector object and we copy its content to another std::vector, but also to a std::list object. Consequently, the content of the std::list object is then copied to a std::set object. For all cases, the std::copy algorithm is used. This algorithm has several arguments: two iterators that define the beginning and end of the source, and an iterator that defines the beginning of the destination. The algorithm copies one element at a time from the input range to the element pointer by the output iterator and then increments the output iterator. Conceptually, it can be implemented as follows:

```
template<typename InputIt, class OutputIt>
OutputIt copy(InputIt first, InputIt last,
              OutputIt d_first)
{
    for (; first != last; (void)++first, (void)++d_first)
    {
        *d_first = *first;
    }
    return d_first;
}
```

> **Important Note**
>
> This algorithm was discussed in *Chapter 5, Type Traits and Conditional Compilation*, when we looked at how its implementation can be optimized with the help of type traits.

Considering the previous example, there are cases when the destination container does not have its content already allocated for the copying to take place. This is the case with copying to the list and to the set. In this case, iterator-like types, std::back_insert_iterator and std:insert_iterator, are used—indirectly through the std::back_inserter and std::inserter helper functions, for inserting elements into a container. The std::back_insert_iterator class uses the push_back function and std::insert_iterator uses the insert function.

There are six iterator categories in C++:

- Input iterator

- Output iterator

- Forward iterator

- Bidirectional iterator

- Random access iterator

- Contiguous iterator

The contiguous iterator category was added in C++17. All operators can be incremented with the prefix or postfix increment operator. The following table shows the additional operations that each category defines:

Category	Properties	Expressions
Input	Increment single-pass	`i++` `++i`
	Equality/inequality comparison	`i == j` `i != j`
	Can be dereferenced (as an rvalue)	`*i` `i->m`
Forward	Increment multi-pass	`i = j; *j++; *i;`
Bidirectional	Decrement	`--i` `i--`

Random-access	Arithmetic operators + and -	`i + n`
		`n + i`
		`i - n`
		`i - j`

	Inequality comparison (with iterators)	`i < j`
		`i < j`
		`i <= j`
		`i >= j`

	Compound assignment	`i += n` `i -= n`
	Offset dereference operator	`i[n]`
Contiguous	Logically adjacent elements are physically adjacent in memory	
Output	Increment single-pass	`i++` `++i`
	Can be dereferenced (as an lvalue)	`*i = v` `*i++ = v`

Table 8.1

With the exception of the output category, each category includes everything about it. That means a forward iterator is an input iterator, a bidirectional iterator is a forward iterator, a random-access iterator is a bidirectional iterator, and, lastly, a contiguous iterator is a random-access iterator. However, iterators in any of the first five categories can also be output iterators at the same time. Such an iterator is called a **mutable** iterator. Otherwise, they are said to be **constant** iterators.

The C++20 standard has added support for concepts and a concepts library. This library defines standard concepts for each of these iterator categories. The following table shows the correlation between them:

Iterator category	Concept
Input iterator	`std::input_iterator`
Forward iterator	`std::output_iterator`
Bidirectional iterator	`std::bidirectional_iterator`
Random-access iterator	`std::random_access_iterator`
Contiguous iterator	`std::contiguous_iterator`
Output iterator	`std::output_iterator`

Table 8.2

> **Important Note**
> Iterator concepts were briefly discussed in *Chapter 6, Concepts and Constraints*.

All containers have the following members:

- `begin`: It returns an iterator to the beginning of the container.
- `end`: It returns an iterator to the end of the container.
- `cbegin`: It returns a constant iterator to the beginning of the container.
- `cend`: It returns a constant iterator to the end of the container.

Some containers also have members that return reverse iterators:

- `rbegin`: It returns a reverse iterator to the beginning of the reversed container.

- `rend`: It returns a reverse iterator to the end of the reversed container.

- `rcbegin`: It returns a constant reverse iterator to the beginning of the reversed container.

- `rcend`: It returns a constant reverse iterator to the end of the reversed container.

There are two things that must be well understood to be able to work with containers and iterators:

- The end of a container is not the last element of the container but the one past the last.

- Reversed iterators provide access to the elements in reverse order. A reversed iterator to the first element of a container is actually the last element of the non-reversed container.

To better understand these two points, let's look at the following example:

```
std::vector<int> v{ 1,2,3,4,5 };

// prints 1 2 3 4 5
std::copy(v.begin(), v.end(),
          std::ostream_iterator<int>(std::cout, " "));

// prints 5 4 3 2 1
std::copy(v.rbegin(), v.rend(),
          std::ostream_iterator<int>(std::cout, " "));
```

The first call to `std::copy` prints the elements of the container in their given order. On the other hand, the second call to `std::copy` prints the elements in their reversed order.

The following diagram illustrates the relationship between iterators and container elements:

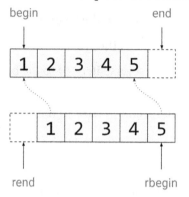

Figure 8.1

A sequence of elements (regardless of what kind of data structure they are stored in memory) delimited by two iterators (a begin and an end, which is the one past the last element) is called a **range**. This term is used extensively in the C++ standard (especially with algorithms) and in literature. It is also the term that gave the name to the ranges library in C++20, which will be discussed in *Chapter 9, The Ranges Library*.

Apart from the set of the begin/end member functions of the standard containers, there are also stand-alone functions with the same name. Their equivalence is presented in the following table:

Member functions	Free functions
c.begin() c.cbegin()	std::begin(c) std::cbegin(c)
c.end() c.cend()	std::end(c) std::cend(c)
c.rbegin() c.rcbegin()	std::rbegin(c) std::rcbegin(c)
c.rend() c.rcend()	std::rend(c) std::rcend(c)

Table 8.3

Although these free functions do not bring much benefit when working with standard containers, they help us in writing generic code that can handle both standard containers and C-like arrays, since all these free functions are overloaded for static arrays. Here is an example:

```
std::vector<int> v{ 1,2,3,4,5 };
std::copy(std::begin(v),  std::end(v),
          std::ostream_iterator<int>(std::cout, " "));

int a[] = { 1,2,3,4,5 };
std::copy(std::begin(a),  std::end(a),
          std::ostream_iterator<int>(std::cout, " "));
```

Without these functions, we would have to write `std::copy(a, a + 5, ...)`. Perhaps a big benefit of these functions is that they enable us to use arrays with range-based for loops, as follows:

```
std::vector<int> v{ 1,2,3,4,5 };
for (auto const& e : v)
   std::cout << e << ' ';

int a[] = { 1,2,3,4,5 };
for (auto const& e : a)
   std::cout << e << ' ';
```

It is not the purpose of this book to teach you how to use each container or the many standard algorithms. However, it should be helpful for you to learn how to create containers, iterators, and algorithms. This is what we will do next.

Creating a custom container and iterator

The best way to understand how containers and iterators work is to experience them first-hand by creating your own. To avoid implementing something that already exists in the standard library, we will consider something different – more precisely, a **circular buffer**. This is a container that, when full, overwrites existing elements. We can think of different ways such a container could work; therefore, it's important that we first define the requirements for it. These are as follows:

- The container should have a fixed capacity that is known at compile-time. Therefore, there would be no runtime memory management.

- The capacity is the number of elements the container can store, and the size is the number of elements it actually contains. When the size equals the capacity, we say the container is full.

- When the container is full, adding a new element will overwrite the oldest element in the container.

- Adding new elements is always done at the end; removing existing elements is always done at the beginning (the oldest element in the container).

- There should be random access to the elements of the container, both with the subscript operator and with iterators.

Based on these requirements, we can think of the following implementation details:

- The elements could be stored in an array. For convenience, this could be the `std::array` class.

- We need two variables, which we call `head` and `tail`, to store the index of the first and last elements of the container. This is needed because due to the circular nature of the container, the beginning and the end shift over time.

- A third variable will store the number of elements in the container. This is useful because otherwise, we would not be able to differentiate whether the container is empty or has one element only from the values of the head and tail indexes.

> **Important Note**
>
> The implementation shown here is provided for teaching purposes only and is not intended as a production-ready solution. The experienced reader will find different aspects of the implementation that could be optimized. However, the purpose here is to learn how to write a container and not how to optimize the implementation.

The following diagram shows a visual representation of such a circular buffer, with a capacity of eight elements with different states:

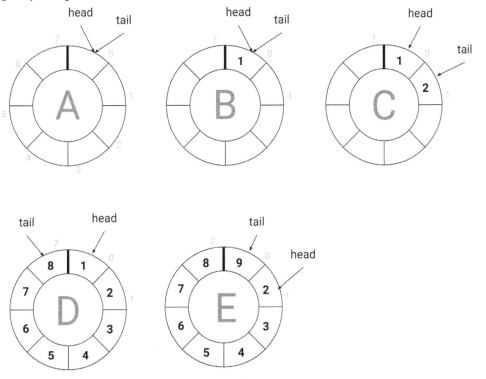

Figure 8.2

What we can see in this diagram is the following:

- **Figure A**: This is an empty buffer. The capacity is 8, the size is 0, and the **head** and **tail** both point to index **0**.
- **Figure B**: This buffer contains one element. The capacity is still 8, the size is 1, and the **head** and **tail** both still point to index **0**.
- **Figure C**: This buffer contains two elements. The size is 2, the **head** contains index **0**, and the **tail** contains index **1**.
- **Figure D**: This buffer is full. The size is 8, which is equal to the capacity, and the **head** contains index **0** and the **tail** contains index **7**.
- **Figure E**: This buffer is still full, but an additional element has been added, triggering the overwriting of the oldest element in the buffer. The size is 8, the **head** contains index **1**, and the **tail** contains index **0**.

Now that we have looked at the semantics of the circular buffer, we can start writing the implementation. We will begin with the container class.

Implementing a circular buffer container

The code for the container class would be too long to be put in a single listing, so we'll break it up into multiple snippets. The first is as follows:

```
template <typename T, std::size_t N>
    requires(N > 0)
class circular_buffer_iterator;


template <typename T, std::size_t N>
    requires(N > 0)
class circular_buffer
{
  // ...
};
```

There are two things here: the forward declaration of a class template called `circular_buffer_iterator`, and a class template called `circular_buffer`. Both have the same template arguments, a type template parameter `T`, representing the type of the elements, and a non-type template parameter, representing the capacity of the buffer. A constraint is used to ensure that the provided value for capacity is always positive. If you are not using C++20, you can replace the constraint with a `static_assert` statement or `enable_if` to enforce the same restriction. The following snippets are all part of the `circular_buffer` class.

First, we have a series of member type definitions that provide aliases to different types that are relevant to the `circular_buffer` class template. These will be used in the implementation of the class. They are shown next:

```
public:
    using value_type = T;
    using size_type = std::size_t;
    using difference_type = std::ptrdiff_t;
    using reference = value_type&;
    using const_reference = value_type const&;
    using pointer = value_type*;
    using const_pointer = value_type const*;
    using iterator = circular_buffer_iterator<T, N>;
    using const_iterator =
        circular_buffer_iterator<T const, N>;
```

Second, we have the data members that store the buffer state. The actual elements are stored in a `std::array` object. The head, tail, and size are all stored in a variable of the `size_type` data type. These members are all private:

```
private:
    std::array<value_type, N> data_;
    size_type                       head_ = 0;
    size_type                       tail_ = 0;
    size_type                       size_ = 0;
```

Third, we have the member functions that implement the functionality described earlier. All the following members are public. The first to list here are the constructors:

```
constexpr circular_buffer() = default;
constexpr circular_buffer(value_type const (&values)[N]) :
    size_(N), tail_(N-1)
{
    std::copy(std::begin(values), std::end(values),
              data_.begin());
}


constexpr circular_buffer(const_reference v):
    size_(N), tail_(N-1)
{
    std::fill(data_.begin(), data_.end(), v);
}
```

There are three constructors defined here (although we can think of additional ones). These are the default constructor (which is also defaulted) that initializes an empty buffer, a constructor from a C-like array of size N, which initializes a full buffer by copying the array elements, and, finally, a constructor that takes a single value and initializes a full buffer by copying that value into each element of the buffer. These constructors allow us to create circular buffers in any of the following ways:

```
circular_buffer<int, 1> b1;                  // {}
circular_buffer<int, 3> b2({ 1, 2, 3 });     // {1, 2, 3}
circular_buffer<int, 3> b3(42);              // {42, 42, 42}
```

Next, we define several member functions that describe the state of the circular buffer:

```cpp
constexpr size_type size() const noexcept
{ return size_; }

constexpr size_type capacity() const noexcept
{ return N; }

constexpr bool empty() const noexcept
{ return size_ == 0; }

constexpr bool full() const noexcept
{ return size_ == N; }

constexpr void clear() noexcept
{ size_ = 0; head_ = 0; tail_ = 0; };
```

The `size` function returns the number of elements in the buffer, the `capacity` function the number of elements that the buffer can hold, the `empty` function to check whether the buffer has no elements (the same as `size() == 0`), and the `full` function to check whether the buffer is full (the same as `size() == N`). There is also a function called `clear` that puts the circular buffer in the empty state. Beware that this function doesn't destroy any element (does not release memory or call destructors) but only resets the values defining the state of the buffer.

We need to access the elements of the buffer; therefore, the following functions are defined for this purpose:

```cpp
constexpr reference operator[](size_type const pos)
{
    return data_[(head_ + pos) % N];
}


constexpr const_reference operator[](size_type const pos) const
{
    return data_[(head_ + pos) % N];
}
```

```cpp
constexpr reference at(size_type const pos)
{
   if (pos < size_)
      return data_[(head_ + pos) % N];

   throw std::out_of_range("Index is out of range");
}

constexpr const_reference at(size_type const pos) const
{
   if (pos < size_)
      return data_[(head_ + pos) % N];

   throw std::out_of_range("Index is out of range");
}

constexpr reference front()
{
   if (size_ > 0) return data_[head_];
   throw std::logic_error("Buffer is empty");
}

constexpr const_reference front() const
{
   if (size_ > 0) return data_[head_];
   throw std::logic_error("Buffer is empty");
}

constexpr reference back()
{
   if (size_ > 0) return data_[tail_];
   throw std::logic_error("Buffer is empty");
}
```

```
constexpr const_reference back() const
{
    if (size_ > 0) return data_[tail_];
    throw std::logic_error("Buffer is empty");
}
```

Each of these members has a const overload that is called for constant instances of the buffer. The constant member returns a constant reference; the non-const member returns a normal reference. These methods are as follows:

- The subscript operator that returns a reference to the element specified by its index, without checking the value of the index

- The at method that works similarly to the subscript operator, except that it checks that the index is smaller than the size and, if not, throws an exception

- The front method that returns a reference to the first element; if the buffer is empty, it throws an exception

- The back method that returns a reference to the last element; if the buffer is empty, it throws an exception

We have members to access the elements, but we need members for adding and removing elements to/from the buffer. Adding new elements always happens at the end, so we'll call this push_back. Removing existing elements always happens at the beginning (the oldest element), so we'll call this pop_front. Let's look first at the former:

```
constexpr void push_back(T const& value)
{
    if (empty())
    {
        data_[tail_] = value;
        size_++;
    }
    else if (!full())
    {
        data_[++tail_] = value;
        size_++;
    }
```

```
    else
    {
        head_ = (head_ + 1) % N;
        tail_ = (tail_ + 1) % N;
        data_[tail_] = value;
    }
}
```

This works based on the defined requirements and the visual representations from *Figure 8.2*:

- If the buffer is empty, copy the value to the element pointed by the tail_ index and increment the size.

- If the buffer is neither empty nor full, do the same but also increment the value of the tail_ index.

- If the buffer is full, increment both the head_ and the tail_ and then copy the value to the element pointed by the tail_ index.

This function copies the value argument to a buffer element. However, this could be optimized for temporaries or objects that are no longer needed after pushing to the buffer. Therefore, an overload that takes an rvalue reference is provided. This moves the value to the buffer, avoiding an unnecessary copy. This overload is shown in the following snippet:

```
constexpr void push_back(T&& value)
{
    if (empty())
    {
        data_[tail_] = value;
        size_++;
    }
    else if (!full())
    {
        data_[++tail_] = std::move(value);
        size_++;
```

```
    }
    else
    {
        head_ = (head_ + 1) % N;
        tail_ = (tail_ + 1) % N;
        data_[tail_] = std::move(value);
    }
}
```

A similar approach is used for implementing the pop_back function to remove elements from the buffer. Here is the implementation:

```
constexpr T pop_front()
{
    if (empty()) throw std::logic_error("Buffer is empty");

    size_type index = head_;

    head_ = (head_ + 1) % N;
    size_--;

    return data_[index];
}
```

This function throws an exception if the buffer is empty. Otherwise, it increments the value of the head_ index and returns the value of the element from the previous position of the head_. This is described visually in the following diagram:

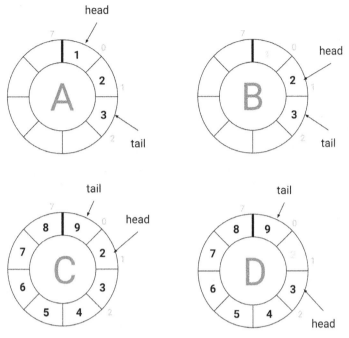

Figure 8.3

What we can see here is the following:

- **Figure A**: The buffer has 3 elements (1, 2, and 3), **head** is at index **0**, and **tail** is at index **2**.

- **Figure B**: An element has been removed from the front, which was index **0**. Therefore, **head** is now index **1** and **tail** is still index **2**. The buffer now has two elements.

- **Figure C**: The buffer has eight elements, which is its maximum capacity, and an element has been overwritten. The **head** is at index **1** and the **tail** is at index **0**.

- **Figure D**: An element has been removed from the front, which was index **1**. The **head** is now at index **2** and the **tail** is still at index **0**. The buffer now has seven elements.

An example of using both the push_back and the pop_front member functions is shown in the next snippet:

```
circular_buffer<int, 4> b({ 1, 2, 3, 4 });
assert(b.size() == 4);


b.push_back(5);
```

```
b.push_back(6);
b.pop_front();

assert(b.size() == 3);
assert(b[0] == 4);
assert(b[1] == 5);
assert(b[2] == 6);
```

Finally, we have the member functions `begin` and `end` that return iterators to the first and one-past-the-last elements of the buffer. Here is their implementation:

```
iterator begin()
{
    return iterator(*this, 0);
}

iterator end()
{
    return iterator(*this, size_);
}

const_iterator begin() const
{
    return const_iterator(*this, 0);
}

const_iterator end() const
{
    return const_iterator(*this, size_);
}
```

To understand these, we need to see how the iterator class is actually implemented. We will explore this in the next section.

Implementing an iterator type for the circular buffer container

We declared the iterator class template at the beginning of the previous section when we started with the `circular_buffer` container. However, we need to define its implementation too. Yet, there is one more thing we must do: in order for the iterator class to be able to access the private members of the container it needs to be declared as a friend. This is done as follows:

```
private:
    friend circular_buffer_iterator<T, N>;
```

Let's look now at the `circular_buffer_iterator` class, which actually has similarities with the container class. This includes the template parameters, the constraints, and the set of member types (some of them being common to those in `circular_buffer`). Here is a snippet of the class:

```
template <typename T, std::size_t N>
requires(N > 0)
class circular_buffer_iterator
{
public:
    using self_type = circular_buffer_iterator<T, N>;
    using value_type = T;
    using reference = value_type&;
    using const_reference = value_type const &;
    using pointer = value_type*;
    using const_pointer = value_type const*;
    using iterator_category =
        std::random_access_iterator_tag;
    using size_type = std::size_t;
    using difference_type = std::ptrdiff_t;

public:
    /* definitions */

private:
    std::reference_wrapper<circular_buffer<T, N>> buffer_;
```

```
    size_type                 index_ = 0;
};
```

The `circular_buffer_iterator` class has a reference to a circular buffer and an index to the element in the buffer that it points to. The reference to `circular_buffer<T, N>` is wrapped inside a `std::reference_wrapper` object. The reason for this will be unveiled shortly. Such an iterator can be explicitly created by providing these two arguments. Therefore, the only constructor looks as follows:

```
explicit circular_buffer_iterator(
   circular_buffer<T, N>& buffer,
   size_type const index):
   buffer_(buffer), index_(index)
{ }
```

If we now look back at the definitions of the `begin` and `end` member functions of `circular_buffer`, we can see that the first argument was `*this`, and the second was 0 for the begin iterator and `size_` for the end iterator. The second value is the offset from the head of the element pointed by the iterator. Therefore, 0 is the first element, and `size_` is the one-past-the-last element in the buffer.

We have decided that we need random access to the elements of the buffer; therefore, the iterator category is random-access. The member type `iterator_category` is an alias for `std::random_access_iterator_tag`. This implies that we need to provide all the operations supported for such an iterator. In the previous section of this chapter, we discussed the iterator categories and the required operations for each category. We will implement all the required ones next.

We start with the requirements for an input iterator, which are as follows:

```
self_type& operator++()
{
   if(index_ >= buffer_.get().size())
      throw std::out_of_range("Iterator cannot be
                    incremented past the end of the range");

   index_++;
   return *this;
}
```

```cpp
self_type operator++(int)
{
    self_type temp = *this;
    ++*this;
    return temp;
}

bool operator==(self_type const& other) const
{
    return compatible(other) && index_ == other.index_;
}

bool operator!=(self_type const& other) const
{
    return !(*this == other);
}

const_reference operator*() const
{
    if (buffer_.get().empty() || !in_bounds())
        throw std::logic_error("Cannot dereferentiate the
                                iterator");
    return buffer_.get().data_[
        (buffer_.get().head_ + index_) %
        buffer_.get().capacity()];
}

const_reference operator->() const
{
    if (buffer_.get().empty() || !in_bounds())
        throw std::logic_error("Cannot dereferentiate the
                                iterator");
    return buffer_.get().data_[
        (buffer_.get().head_ + index_) %
        buffer_.get().capacity()];
}
```

We have implemented here incrementing (both pre- and post-fix), checking for equality/ inequality, and dereferencing. The * and -> operators throw an exception if the element cannot be dereferenced. The cases when this happens are when the buffer is empty, or the index is not within bounds (between `head_` and `tail_`). We used two helper functions (both private), called `compatible` and `is_bounds`. These are shown next:

```
bool compatible(self_type const& other) const
{
    return buffer_.get().data_.data() ==
            other.buffer_.get().data_.data();
}


bool in_bounds() const
{
    return
        !buffer_.get().empty() &&
        (buffer_.get().head_ + index_) %
        buffer_.get().capacity() <= buffer_.get().tail_;
}
```

A **forward iterator** is an input iterator that is also an **output iterator**. The requirements for output iterators will be discussed toward the end of this section. Those for the input iterators we have seen previously. Apart from those, forward iterators can be used in multi-pass algorithms, which is possible because performing an operation on a dereferenceable forward iterator does not make its iterator value non-dereferenceable. That means if a and b are two forward iterators and they are equal, then either they are non-dereferenceable, otherwise, their iterator values, *a and *b, refer to the same object. The opposite is also true, meaning that if *a and *b are equal, then a and b are also equal. This is true for our implementation.

The other requirement for forward iterators is that they are swappable. That means if a and b are two forward iterators, then swap(a, b) should be a valid operation. This leads us back to using a `std::reference_wrapper` object to hold a reference to a `circular_buffer<T, N>`. References are not swappable, which would have made `circular_buffer_iterator` not swappable either. However, `std::reference_wrapper` is swappable, and that also makes our iterator type swappable. That can be verified with a `static_assert` statement, such as the following:

```
static_assert(
    std::is_swappable_v<circular_buffer_iterator<int, 10>>);
```

> **Important Note**
> An alternative to using `std::reference_wrapper` is to use a raw pointer to a `circular_buffer` class, since pointers can be assigned values and are, therefore swappable. It is a matter of style and personal choice which one to use. In this example, I preferred the solution that avoided the raw pointer.

For fulfilling the requirements for the bidirectional iterator category, we need to support decrementing. In the next snippet, you can see the implementation of both pre- and post-fix decrement operators:

```cpp
self_type& operator--()
{
    if(index_ <= 0)
        throw std::out_of_range("Iterator cannot be
                decremented before the beginning of the range");

    index_--;
    return *this;
}

self_type operator--(int)
{
    self_type temp = *this;
    --*this;
    return temp;
}
```

Finally, we have the requirements of the **random-access iterators**. The first requirements to implement are arithmetic (+ and -) and compound (+= and -=) operations. These are shown next:

```cpp
self_type operator+(difference_type offset) const
{
    self_type temp = *this;
    return temp += offset;
}

self_type operator-(difference_type offset) const
```

```
{
    self_type temp = *this;
    return temp -= offset;
}


difference_type operator-(self_type const& other) const
{
    return index_ - other.index_;
}


self_type& operator +=(difference_type const offset)
{
    difference_type next =
        (index_ + next) % buffer_.get().capacity();
    if (next >= buffer_.get().size())
        throw std::out_of_range("Iterator cannot be
                    incremented past the bounds of the range");

    index_ = next;
    return *this;
}


self_type& operator -=(difference_type const offset)
{
    return *this += -offset;
}
```

Random-access iterators must support inequality comparison with other operations. That means, we need to overload the <, <=, >, and >= operators. However, the <=, >, and >= operators can be implemented based on the < operator. Therefore, their definition can be as follows:

```
bool operator<(self_type const& other) const
{
    return index_ < other.index_;
}
```

```cpp
bool operator>(self_type const& other) const
{
    return other < *this;
}

bool operator<=(self_type const& other) const
{
    return !(other < *this);
}

bool operator>=(self_type const& other) const
{
    return !(*this < other);
}
```

Last, but not least, we need to provide access to elements with the subscript operator ([ ]).
A possible implementation is the following:

```cpp
value_type& operator[](difference_type const offset)
{
    return *((*this + offset));
}

value_type const & operator[](difference_type const offset)
const
{
    return *((*this + offset));
}
```

With this, we have completed the implementation of the iterator type for the circular
buffer. If you had trouble following the multitude of code snippets for these two classes,
you can find the complete implementation in the GitHub repository for the book.
A simple example for using the iterator type is shown next:

```cpp
circular_buffer<int, 3> b({1, 2, 3});
std::vector<int> v;
for (auto it = b.begin(); it != b.end(); ++it)
{
```

```
        v.push_back(*it);
    }
```

This code can be actually simplified with range-based for loops. In this case, we don't use iterators directly, but the compiler-generated code does. Therefore, the following snippet is equivalent to the previous one:

```
circular_buffer<int, 3> b({ 1, 2, 3 });
std::vector<int> v;
for (auto const e : b)
{
    v.push_back(e);
}
```

However, the implementation provided here for `circular_buffer_iterator` does not allow the following piece of code to compile:

```
circular_buffer<int, 3> b({ 1,2,3 });
*b.begin() = 0;

assert(b.front() == 0);
```

This requires that we are able to write elements through iterators. However, our implementation doesn't meet the requirements for the output iterator category. This requires that expressions such as `*it = v`, or `*it++ = v` are valid. To do so, we need to provide non-const overloads of the `*` and `->` operators that return non-const reference types. This can be done as follows:

```
reference operator*()
{
    if (buffer_.get().empty() || !in_bounds())
        throw std::logic_error("Cannot dereferentiate the
                                iterator");
    return buffer_.get().data_[
        (buffer_.get().head_ + index_) %
        buffer_.get().capacity()];
}

reference operator->()
{
```

```
    if (buffer_.get().empty() || !in_bounds())
        throw std::logic_error("Cannot dereferentiate the
                               iterator");
    return buffer_.get().data_[
        (buffer_.get().head_ + index_) %
        buffer_.get().capacity()];
}
```

More examples for using the `circular_buffer` class with and without iterators can be found in the GitHub repository. Next, we will focus our attention on implementing a general-purpose algorithm that works for any range, including the `circular_buffer` container we defined here.

Writing a custom general-purpose algorithm

In the first section of this chapter, we saw why abstracting access to container elements with iterators is key for building general-purpose algorithms. However, it should be useful for you to practice writing such an algorithm because it can help you better understand the use of iterators. Therefore, in this section, we will write a general-purpose algorithm.

The standard library features many such algorithms. One that is missing is a **zipping algorithm**. What zipping means is actually interpreted or understood differently by different people. For some, zipping means taking two or more input ranges and creating a new range with the elements from the input ranges intercalated. This is exemplified in the following diagram:

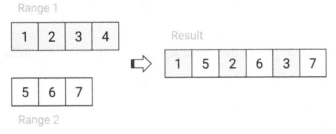

Figure 8.4

For others, zipping means taking two or more input ranges and creating a new range, with elements being tuples formed from the elements of the input ranges. This is shown in the next diagram:

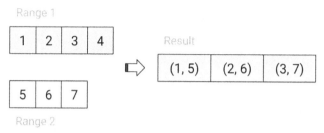

Figure 8.5

In this section, we will implement the first algorithm. In order to avoid confusion, we will call this **flatzip**. Here are the requirements for it:

- The algorithm takes two input ranges and writes to an output range.

- The algorithm takes iterators as arguments. A pair of first and last input iterators define the bounds of each input range. An output iterator defines the beginning of the output range where the elements will be written to.

- The two input ranges should contain elements of the same type. The output range must have elements of the same type or a type to which the input type is implicitly convertible.

- If the two input ranges are of different sizes, the algorithm stops when the smallest of the two has been processed (as shown in the previous diagrams).

- The return value is the output iterator to the one-past-the-last element that was copied.

A possible implementation for the described algorithm is shown in the following listing:

```
template <typename InputIt1, typename InputIt2,
          typename OutputIt>
OutputIt flatzip(
   InputIt1 first1, InputIt1 last1,
   InputIt2 first2, InputIt2 last2,
   OutputIt dest)
{
   auto it1 = first1;
   auto it2 = first2;
```

```
    while (it1 != last1 && it2 != last2)
    {
        *dest++ = *it1++;
        *dest++ = *it2++;
    }


    return dest;
}
```

As you can see in the snippet, the implementation is quite simple. All we do here is iterate through both input ranges at the same time and copy elements alternately from them to the destination range. The iteration on both input ranges stops when the end of the smallest range is reached. We can use this algorithm as follows:

```
// one range is empty
std::vector<int> v1 {1,2,3};
std::vector<int> v2;
std::vector<int> v3;

flatzip(v1.begin(), v1.end(), v2.begin(), v2.end(),
        std::back_inserter(v3));
assert(v3.empty());

// neither range is empty
std::vector<int> v1 {1, 2, 3};
std::vector<int> v2 {4, 5};
std::vector<int> v3;

flatzip(v1.begin(), v1.end(), v2.begin(), v2.end(),
        std::back_inserter(v3));
assert(v3 == std::vector<int>({ 1, 4, 2, 5 }));
```

These examples use `std::vector` for both the input and output ranges. However, the `flatzip` algorithm knows nothing about containers. The elements of the container are accessed with the help of iterators. Therefore, as long as the iterators meet the specified requirements, we can use any container. This includes the `circular_buffer` container we previously wrote since `circular_buffer_container` meets the requirements for both the input and output iterator categories. This means we can also write the following snippet:

```
circular_buffer<int, 4> a({1, 2, 3, 4});
circular_buffer<int, 3> b({5, 6, 7});
circular_buffer<int, 8> c(0);


flatzip(a.begin(), a.end(), b.begin(), b.end(), c.begin());


std::vector<int> v;
for (auto e : c)
    v.push_back(e);
assert(v == std::vector<int>({ 1, 5, 2, 6, 3, 7, 0, 0 }));
```

We have two input circular buffers: a, which has four elements, and b, which has three elements. The destination circular buffer has a capacity of eight elements, all initialized with zero. After applying the flatzip algorithm, six elements of the destination circular buffer will be written with values from the a and b buffers. The result is that the circular buffer will contain the elements 1, 5, 2, 6, 3, 7, 0, 0.

Summary

This chapter was dedicated to seeing how templates can be used to build general-purpose libraries. Although we couldn't cover these topics in great detail, we have explored the design of containers, iterators, and algorithms from the C++ standard library. These are the pillars of the standard library. We spent most of the chapter understanding what it takes to write a container similar to the standard ones as well as an iterator class to provide access to its elements. For this purpose, we implemented a class that represents a circular buffer, a data structure of fixed size where elements are overwritten once the container is full. Lastly, we implemented a general-purpose algorithm that zips elements from two ranges. This works for any container including the circular buffer container.

Ranges, as discussed in this chapter, are an abstract concept. However, that changed with C++20, which introduced a more concrete concept of ranges with the new ranges library. This is what we will discuss in the final chapter of this book.

Questions

1. What are the sequence containers from the standard library?

2. What are the common member functions defined in the standard containers?

3. What are iterators and how many categories exist?

4. What operations does a random-access iterator support?

5. What are range access functions?

9
The Ranges Library

The previous chapter was dedicated to understanding the three main pillars of the standard library: containers, iterators, and algorithms. Throughout that chapter, we used the abstract concept of range to represent a sequence of elements delimited by two iterators. The C++20 standard makes it easier to work with ranges by providing a ranges library, consisting of two main parts: on one hand, types that define non-owning ranges and adaptations of ranges, and on the other hand, algorithms that work with these range types and do not require iterators to define a range of elements.

In this final chapter, we will address the following topics:

- Moving from abstract ranges to the ranges library

- Understanding range concepts and views

- Understanding the constrained algorithms

- Writing your own range adaptor

By the end of this chapter, you will have a good understanding of the content of the ranges library and you will be able to write your own range adaptor.

Let's begin the chapter with a transition from the abstract concept of a range to the C++20 ranges library.

Advancing from abstract ranges to the ranges library

We have used the term *range* many times in the previous chapter. A **range** is an abstraction of a sequence of elements, delimited by two iterators (one to the first element of the sequence, and one to the one-past-the-last element). Containers such as `std::vector`, `std::list`, and `std::map` are concrete implementations of the range abstraction. They have ownership of the elements and they are implemented using various data structures, such as arrays, linked-lists, or trees. The standard algorithms are generic. They are container-agnostic. They know nothing about `std::vector`, `std::list`, or `std::map`. They handle range abstractions with the help of iterators. However, this has a shortcoming: we always need to retrieve a beginning and end iterator from a container. Here are some examples:

```cpp
// sorts a vector
std::vector<int> v{ 1, 5, 3, 2, 4 };
std::sort(v.begin(), v.end());


// counts even numbers in an array
std::array<int, 5> a{ 1, 5, 3, 2, 4 };
auto even = std::count_if(
    a.begin(), a.end(),
    [](int const n) {return n % 2 == 0; });
```

There are few cases when you need to process only a part of the container's elements. In the vast majority of cases, you just have to write `v.begin()` and `v.end()` over and over again. This includes variations such as calls to `cbegin()`/`cend()`, `rbegin()`/`rend()`, or the stand-alone functions `std::begin()`/`std::end()`, and so on. Ideally, we would prefer to shorten all this and be able to write the following:

```cpp
// sorts a vector
std::vector<int> v{ 1, 5, 3, 2, 4 };
sort(v);


// counts even numbers in an array
std::array<int, 5> a{ 1, 5, 3, 2, 4 };
auto even = std::count_if(
    a,
    [](int const n) {return n % 2 == 0; });
```

On the other hand, we often need to compose operations. Most of the time that involves many operations and code that is too verbose even when using standard algorithms. Let's consider the following example: given a sequence of integers, we want to print to the console the square of all even numbers, except the first two, in descending order of their value (not their position in the sequence). There are multiple ways to solve the problem. The following is a possible solution:

```cpp
std::vector<int> v{ 1, 5, 3, 2, 8, 7, 6, 4 };

// copy only the even elements
std::vector<int> temp;
std::copy_if(v.begin(), v.end(),
             std::back_inserter(temp),
             [](int const n) {return n % 2 == 0; });

// sort the sequence
std::sort(temp.begin(), temp.end(),
          [](int const a, int const b) {return a > b; });

// remove the first two
temp.erase(temp.begin() + temp.size() - 2, temp.end());

// transform the elements
std::transform(temp.begin(), temp.end(),
               temp.begin(),
               [](int const n) {return n * n; });

// print each element
std::for_each(temp.begin(), temp.end(),
              [](int const n) {std::cout << n << '\n'; });
```

I believe most people would agree that, although anyone familiar with the standard algorithms can easily read this code, it's still a lot to write. It also requires a temporary container and repetitive calls to begin/end. Therefore, I also expect most people would more easily understand the following version of the previous code, and probably prefer to write it as such:

```cpp
std::vector<int> v{ 1, 5, 3, 2, 8, 7, 6, 4 };
sort(v);
auto r = v
         | filter([](int const n) {return n % 2 == 0; })
         | drop(2)
         | reverse
         | transform([](int const n) {return n * n; });

for_each(r, [](int const n) {std::cout << n << '\n'; });
```

This is what the C++20 standard provides with the help of the ranges library. This has two main components:

- **Views** or **range adaptors**, which represent non-owning iterable sequences. They enable us to compose operations more easily such as in the last example.

- **Constrained algorithms**, which enable us to operate on concrete ranges (standard containers or ranges) and not on abstract ranges delimited with a pair of iterators (although that's possible too).

We will explore these two offerings of the ranges library in the next sections, and we will begin with ranges.

Understanding range concepts and views

The term *range* refers to an abstraction that defines a sequence of elements bounded by start and end iterators. A range, therefore, represents an iterable sequence of elements. However, such a sequence can be defined in several ways:

- With a begin iterator and an end sentinel. Such a sequence is iterated from the beginning to the end. A **sentinel** is an object that indicates the end of the sequence. It can have the same type as the iterator type or it can be of a different type.

- With a start object and a size (number of elements), representing a so-called counted sequence. Such a sequence is iterated N times (where N represents the size) from the start.

- With a start and a predicate, representing a so-called conditionally terminated sequence. Such a sequence is iterated from the start until the predicate returns false.

- With only a start value, representing a so-called unbounded sequence. Such a sequence can be iterated indefinitely.

All these kinds of iterable sequences are considered ranges. Because a range is an abstraction, the C++20 library defines a series of concepts to describe requirements for range types. These are available in the `<ranges>` header and the `std::ranges` namespace. The following table presents the list of range concepts:

Name	Description
range	Defines the requirements for a type R to be a range by providing a begin iterator and an end sentinel. The iterator and sentinel can be of different types.
borrowed_range	Defines the requirements for a type R so that a function can take an object of this type by value and return iterators, obtained from this object without the danger of dangling.
sized_range	Defines the requirements for a type R to be a range that knows its size in constant time.
common_range	Defines the requirements for a type R to be a range whose iterator and sentinel types are identical.
view	Defines the requirements for a type R that is a range to have constant time copy, move, and assignment operations.
viewable_range	Defines the requirements for a range type R to be convertible to a view.
input_range	Requires that a range type has the iterator type satisfy the input_iterator concept.
output_range	Requires that a range type has the iterator type satisfy the output_iterator concept.
forward_range	Requires that a range type has the iterator type satisfy the forward_iterator concept.
bidirectional_range	Requires that a range type has the iterator type satisfy the bidirectional_iterator concept.
random_access_range	Requires that a range type has the iterator type satisfy the random_access_iterator concept.
contiguous_range	Requires that a range type has the iterator type satisfy the contiguous_iterator concept.

Table 9.1

The standard library defines a series of access functions for containers and arrays. These include `std::begin` and `std::end` instead of member functions `begin` and `end`, `std::size` instead of member function `size`, and so on. These are called **range access functions**. Similarly, the ranges library defines a set of range access functions. These are designed for ranges and are available in the `<ranges>` and `<iterator>` headers and the `std::ranges` namespace. They are listed in the next table:

Range access for ranges	Equivalent range access for containers/arrays	Description
begin / end cbegin / cend	begin / end cbegin / cend	Returns an iterator and constant iterator respectively to the beginning/end of a range.
rbegin / rend crbegin / crend	rbegin / rend crbegin / crend	Returns a reverse iterator and constant reverse iterator respectively, to the beginning / end of a range.
size / ssize	size / ssize	Returns the size of a range as an integral or signed integral value.
empty	empty	Returns a Boolean value, indicating whether the range is empty.
data / cdata	data	Returns a pointer to the beginning of a contiguous range and a read-only contiguous range respectively.

Table 9.2

The use of some of these functions is demonstrated in the following snippet:

```
std::vector<int> v{ 8, 5, 3, 2, 4, 7, 6, 1 };
auto r = std::views::iota(1, 10);

std::cout << "size(v)=" << std::ranges::size(v) << '\n';
std::cout << "size(r)=" << std::ranges::size(r) << '\n';

std::cout << "empty(v)=" << std::ranges::empty(v) << '\n';
std::cout << "empty(r)=" << std::ranges::empty(r) << '\n';
```

```
std::cout << "first(v)=" << *std::ranges::begin(v) << '\n';
std::cout << "first(r)=" << *std::ranges::begin(r) << '\n';

std::cout << "rbegin(v)=" << *std::ranges::rbegin(v)
          << '\n';
std::cout << "rbegin(r)=" << *std::ranges::rbegin(r)
          << '\n';

std::cout << "data(v)=" << *std::ranges::data(v) << '\n';
```

In this snippet, we used a type called `std::views::iota`. As the namespace implies, this is a view. A **view** is a range with additional restrictions. Views are lightweight objects with non-owning semantics. They present a view of an underlying sequence of elements (a range) in a way that does not require copying or mutating the sequence. The key feature is lazy evaluation. That means that regardless of the transformation they apply, they perform it only when an element is requested (iterated) and not when the view is created.

There is a series of views provided with C++20, and new views have been also included in C++23. Views are available in the `<ranges>` header and `std::ranges` namespace in the form, `std::ranges::abc_view`, such as `std::ranges::iota_view`. However, for convenience of use, in the `std::views` namespace, a variable template of the form, `std::views::abc`, such as `std::views::iota`, also exists. This is what we saw in the previous example. Here are two equivalent examples for using `iota`:

```
// using the iota_view type
for (auto i : std::ranges::iota_view(1, 10))
    std::cout << i << '\n';

// using the iota variable template
for (auto i : std::views::iota(1, 10))
    std::cout << i << '\n';
```

The `iota` view is part of a special category of views called **factories**. These factories are views over newly generated ranges. The following factories are available in the ranges library:

Type	Variable	Description
`ranges::empty_view`	`ranges::views::empty`	Generates a view with no elements of a T type.
`ranges::single_view`	`ranges::views::single`	Generates a view with a single element of a T type.
`ranges::iota_view`	`ranges::views::iota`	Generates a view of a sequence of consecutive elements, from a start value to an end value (a bounded view) or indefinitely (an unbounded view).
`ranges::basic_iostream_view`	`ranges::views::istream`	Generates a view of a sequence of elements by applying the operator `>>` repeatedly.

Table 9.3

If you are wondering why `empty_view` and `single_view` are useful, the answer should not be hard to find. These are useful in template code that handles ranges where empty ranges or ranges with one element are valid inputs. You don't want multiple overloads of a function template for handling these special cases; instead, you can pass an `empty_view` or `single_view` range. The following snippets show several examples of using these factories. These snippets should be self-explanatory:

```
constexpr std::ranges::empty_view<int> ev;
static_assert(std::ranges::empty(ev));
static_assert(std::ranges::size(ev) == 0);
static_assert(std::ranges::data(ev) == nullptr);


constexpr std::ranges::single_view<int> sv{42};
static_assert(!std::ranges::empty(sv));
static_assert(std::ranges::size(sv) == 1);
static_assert(*std::ranges::data(sv) == 42);
```

For `iota_view`, we have already seen a couple of examples with a bounded view. The next snippet shows again an example not only using a bounded view generated with `iota` but also an unbounded view, also generated with `iota`:

```
auto v1 = std::ranges::views::iota(1, 10);
std::ranges::for_each(
        v1,
        [](int const n) {std::cout << n << '\n'; });

auto v2 = std::ranges::views::iota(1) |
            std::ranges::views::take(9);
std::ranges::for_each(
        v2,
        [](int const n) {std::cout << n << '\n'; });
```

This last example utilizes another view called `take_view`. This produces a view of the first *N* elements (in our example, 9) of another view (in our case, the unbounded view produced with `iota`). We will discuss more about this shortly. But first, let's take an example using the fourth view factory, `basic_iostream_view`. Let's consider we have a list of article prices in a text, separated by a space. We need to print the total sum of these prices. There are different ways to solve it, but a possible solution is given here:

```
auto text = "19.99 7.50 49.19 20 12.34";
auto stream = std::istringstream{ text };
std::vector<double> prices;

double price;
while (stream >> price)
{
    prices.push_back(price);
}

auto total = std::accumulate(prices.begin(), prices.end(),
                        0.0);
std::cout << std::format("total: {}\n", total);
```

The highlighted part can be replaced with the following two lines of code that use `basic_iostream_view` or, more precisely, the `istream_view` alias template:

```
for (double const price :
        std::ranges::istream_view<double>(stream))
{
    prices.push_back(price);
}
```

What the `istream_view` range factory is doing is applying the operator `>>` repeatedly on the `istringstream` object and producing a value each time it is applied. You cannot specify a delimiter; it only works with whitespaces. If you prefer to use standard algorithms rather than handcrafted loops, you can use the `ranges::for_each` constrained algorithm to produce the same result, as follows:

```
std::ranges::for_each(
    std::ranges::istream_view<double>(stream),
    [&prices](double const price) {
        prices.push_back(price); });
```

The examples given so far in this chapter included views such as `filter`, `take`, `drop`, and `reverse`. These are just a few of the standard views available in C++20. More are being added to C++23, and probably even more to future standard versions. The entire set of standard views is listed in the following table:

Type (in namespace ranges)	Variable (in namespace ranges::view)	C++ version	Description
`filter_view`	`filter`	C++20	A type that represents a range adaptor that provides a view of an underlying range, which only includes the elements that satisfy a predicate.
`transform_view`	`transform`	C++20	A type that represents a range adaptor that provides a view of an underlying range, with a transformation applied to each element of the range.

Type (in namespace ranges)	Variable (in namespace ranges::view)	C++ version	Description
`take_view`	`take`	C++20	A type that represents a range adaptor that provides a view to the first N elements of an underlying sequence.
`take_while_view`	`take_while`	C++20	A type that represents a range adaptor that provides a view of the elements of an underlying sequence, starting with the beginning and until the first element that no longer satisfies a specified predicate.
`drop_view`	`drop`	C++20	A type that represents a range adaptor that provides a view of the elements of underlying sequence with the exception of the first N elements (which are skipped).
`drop_while_view`	`drop_while`	C++20	A type that represents a range adaptor that provides a view of the elements of an underlying sequence, starting with the first element that does not satisfy a given predicate.
`join_view`	`join`	C++20	A type that represents a range adaptor that provides a view of a sequence produced from flattening multiple ranges.
`join_with_view`	`join_with`	C++23	A type that represents a range adaptor that provides a view of a sequence produced from flattening multiple ranges, with a specified delimiter inserted between the elements of the view.

Type (in namespace ranges)	Variable (in namespace ranges::view)	C++ version	Description
split_view	split	C++20	A type that represents a range adaptor that provides a view of a sequence of ranges produced from splitting a range on a specified delimited. The range cannot be an input range and the lazy semantics of views are not observed.
lazy_split_view	lazy_split	C++20	Same as split_view, except that it also works with input ranges and observes the lazy mechanism of ranges.
reverse_view	reverse	C++20	A type that represents a range adaptor that provides a view of the elements of an underlying range in reverse order.
keys_view	keys	C++20	A type that represents a range adaptor that provides a view projected from the first element of tuple-like values (std::pair and std::tuple) of an underlying view.
values_view	values	C++20	A type that represents a range adaptor that provides a view projected from the second element of tuple-like values (std::pair and std::tuple) of an underlying view.
elements_view	elements	C++20	A type that represents a range adaptor that provides a view projected from the Nth element of tuple-like values of an underlying view.
zip_view	zip	C++23	A type that represents a range adaptor that provides a view built from one or more underlying views, projecting the Nth element from each view into a tuple.

Type (in namespace ranges)	Variable (in namespace ranges::view)	C++ version	Description
`zip_ transform_ view`	`zip_ transform`	C++23	A type that represents a range adaptor that provides a view built from one or more underlying views and an invocable object, whose elements are computed by applying the invocable object to the *Nth* element of each underlying view.
`adjacent_ view`	`adjacent`	C++23	A type that represents a range adaptor that provides a view of tuple-like values, projected by taking N consecutive elements of an underlying view.
`adjacent_ transform_ view`	`adjacent_ transform`	C++23	A type that represents a range adaptor that provides a view of values projected by applying an invocable object to N consecutive elements of an underlying view.

Table 9.4

Apart from the views (range adaptors) listed in the previous table, there are a few more that can be useful in some particular scenarios. For completeness, these are listed in the next table:

Type (in namespace ranges)	Variable (in namespace ranges::view)	C++ version	Description
	`all`	C++20	An object that creates a view that includes all the elements of the range argument.
	`all_t`	C++20	An alias template for the view type of a range that can be safely converted to a view.

Type (in namespace ranges)	Variable (in namespace ranges::view)	C++ version	Description
	`counted`	C++20	An object that creates a view with N elements of a range, starting with the one represented by a given iterator.
`ref_view`		C++20	A view type that wraps a reference to another range.
`owning_view`		C++20	A view type that stores a given range. It has sole ownership of the stored range and move-only semantics.
`common_view`	`common`	C++20	A type that adapts a view with different types for the iterator and sentinel pair, with a view that uses the same type for both iterator and sentinel types.

Table 9.5

Now that we have enumerated all the standard range adaptors, let's take a look at more examples using some of them.

Exploring more examples

Previously, in this section, we saw the following example (this time with explicit namespaces):

```
namespace rv = std::ranges::views;
std::ranges::sort(v);
auto r = v
        | rv::filter([] (int const n) {return n % 2 == 0; })
        | rv::drop(2)
        | rv::reverse
        | rv::transform([] (int const n) {return n * n; });
```

This is actually the shorter and more readable version of the following:

```
std::ranges::sort(v);auto r =
  rv::transform(
```

```
rv::reverse(
  rv::drop(
    rv::filter(
      v,
      [](int const n) {return n % 2 == 0; }),
    2)),
  [](int const n) {return n * n; });
```

The first version is possible because the pipe operator (|) is overloaded to simplify the composition of views in a more human-readable form. Some range adaptors take one argument, and some may take multiple arguments. The following rules apply:

- If a range adaptor A takes one argument, a view V, then A(V) and V|A are equivalent. Such a range adaptor is reverse_view, and an example is shown here:

```
std::vector<int> v{ 1, 5, 3, 2, 8, 7, 6, 4 };
namespace rv = std::ranges::views;
auto r1 = rv::reverse(v);
auto r2 = v | rv::reverse;
```

- If a range adaptor A takes multiple arguments, a view V and args..., then A(V, args...), A(args...)(V), and V|A(args...) are equivalent. Such a range adaptor is take_view, and an example is shown here:

```
std::vector<int> v{ 1, 5, 3, 2, 8, 7, 6, 4 };
namespace rv = std::ranges::views;
auto r1 = rv::take(v, 2);
auto r2 = rv::take(2)(v);
auto r3 = v | rv::take(2);
```

So far, we have seen the likes of filter, transform, reverse, and drop put to use. To complete this part of the chapter, let's take a series of examples to demonstrate the use of the views from *Table 8.7*. In all the following examples, we will consider rv as an alias for the std::ranges::views namespace:

- Print the last two odd numbers from a sequence, in reverse order:

```
std::vector<int> v{ 1, 5, 3, 2, 4, 7, 6, 8 };

for (auto i : v |
  rv::reverse |
```

```
    rv::filter([](int const n) {return n % 2 == 1; }) |
    rv::take(2))
{
    std::cout << i << '\n'; // prints 7 and 3
}
```

- Print the subsequence of consecutive numbers smaller than 10 from a range that does not include the first consecutive odd numbers:

```
std::vector<int> v{ 1, 5, 3, 2, 4, 7, 16, 8 };
for (auto i : v |
  rv::take_while([](int const n){return n < 10; }) |
  rv::drop_while([](int const n){return n % 2 == 1; })
)
{
    std::cout << i << '\n'; // prints 2 4 7
}
```

- Print the first elements, the second elements, and, respectively, the third elements from a sequence of tuples:

```
std::vector<std::tuple<int,double,std::string>> v =
{
    {1, 1.1, "one"},
    {2, 2.2, "two"},
    {3, 3.3, "three"}
};

for (auto i : v | rv::keys)
    std::cout << i << '\n'; // prints 1 2 3

for (auto i : v | rv::values)
    std::cout << i << '\n'; // prints 1.1 2.2 3.3

for (auto i : v | rv::elements<2>)
    std::cout << i << '\n'; // prints one two three
```

- Print all the elements from a vector of vectors of integers:

```cpp
std::vector<std::vector<int>> v {
    {1,2,3}, {4}, {5, 6}
};
for (int const i : v | rv::join)
    std::cout << i << ' ';   // prints 1 2 3 4 5 6
```

- Print all the elements from a vector of vectors of integers but insert a 0 between the elements of each vector. The range adaptor join_with is new to C++23 and may not be supported yet by compilers:

```cpp
std::vector<std::vector<int>> v{
    {1,2,3}, {4}, {5, 6}
};
for(int const i : v | rv::join_with(0))
    std::cout << i << ' ';   // print 1 2 3 0 4 0 5 6
```

- Print the individual words from a sentence, where the delimited is a space:

```cpp
std::string text{ "this is a demo!" };
constexpr std::string_view delim{ " " };
for (auto const word : text | rv::split(delim))
{
    std::cout << std::string_view(word.begin(),
                                  word.end())
              << '\n';
}
```

- Create a view of tuples from the elements of an array of integers and a vector of doubles:

```cpp
std::array<int, 4> a {1, 2, 3, 4};
std::vector<double> v {10.0, 20.0, 30.0};

auto z = rv::zip(a, v)
// { {1, 10.0}, {2, 20.0}, {3, 30.0} }
```

- Create a view with the multiplied elements of an array of integers and a vector of doubles:

```
std::array<int, 4> a {1, 2, 3, 4};
std::vector<double> v {10.0, 20.0, 30.0};

auto z = rv::zip_transform(
    std::multiplies<double>(), a, v)
// { {1, 10.0}, {2, 20.0}, {3, 30.0} }
```

- Print the pairs of adjacent elements of a sequence of integers:

```
std::vector<int> v {1, 2, 3, 4};
for (auto i : v | rv::adjacent<2>)
{
    // prints: (1, 2) (2, 3) (3, 4)
    std::cout << std::format("({},{})",
                        i.first, i.second)";
}
```

- Print the values obtained from multiplying each three consecutive values from a sequence of integers:

```
std::vector<int> v {1, 2, 3, 4, 5};
for (auto i : v | rv::adjacent_transform<3>(
    std::multiplies()))
{
    std::cout << i << ' '; // prints: 3 24 60
}
```

These examples will hopefully help you understand the possible use cases for each of the available views. You can find more examples in the source code accompanying the book, as well as in the articles mentioned in the *Further reading* section. In the next section, we will discuss the other part of the ranges library, the constrained algorithms.

Understanding the constrained algorithms

The standard library provides over one hundred general-purpose algorithms. As we discussed in the introductory section for the ranges library earlier, these have one thing in common: they work with abstract ranges with the help of iterators. They take iterators as arguments and they sometimes return iterators. That makes it cumbersome to repeatedly use with standard containers or arrays. Here is an example:

```cpp
auto l_odd = [](int const n) {return n % 2 == 1; };

std::vector<int> v{ 1, 1, 2, 3, 5, 8, 13 };
std::vector<int> o;
auto e1 = std::copy_if(v.begin(), v.end(),
                       std::back_inserter(o),
                       l_odd);

int arr[] = { 1, 1, 2, 3, 5, 8, 13 };
auto e2 = std::copy_if(std::begin(arr), std::end(arr),
                       std::back_inserter(o),
                       l_odd);
```

In this snippet, we have a vector v and an array arr, and we copy the odd elements from each of these two to a second vector, o. For this, the std::copy_if algorithm is used. This takes begin and end input iterators (defining the input range), an output iterator to a second range, where the copied elements will be inserted, and a unary predicate (in this example, a lambda expression). What it returns is an iterator to the destination range past the last copied element.

If we look at the declaration of the std::copy_if algorithm, we will find the following two overloads:

```cpp
template <typename InputIt, typename OutputIt,
          typename UnaryPredicate>
constexpr OutputIt copy_if(InputIt first, InputIt last,
                           OutputIt d_first,
                           UnaryPredicate pred);

template <typename ExecutionPolicy,
          typename ForwardIt1, typename ForwardIt2,
          typename UnaryPredicate>
```

```
ForwardIt2 copy_if(ExecutionPolicy&& policy,
                   ForwardIt1 first, ForwardIt1 last,
                   ForwardIt2 d_first,
                   UnaryPredicate pred);
```

The first overload is the one used and described here. The second overload was introduced in C++17. This allows you to specify an execution policy such as parallel or sequential. This basically enables the parallel execution of the standard algorithms. However, this is not relevant to the topic of this chapter, and we will not explore it further.

Most of the standard algorithms have a new constrained version in the `std::ranges` namespace. These algorithms are found in the `<algorithm>`, `<numeric>`, and `<memory>` headers and have the following traits:

- They have the same name as the existing algorithms.

- They have overloads that allow you to specify a range, either with a begin iterator and an end sentinel, or as a single range argument.

- They have modified return types that provide more information about the execution.

- They support projections to apply to the processed elements. A **projection** is an entity that can be invoked. It can be a pointer to a member, a lambda expression, or a function pointer. Such a projection is applied to the range element *before* the algorithm logic uses the element.

Here is how the overloads of the `std::ranges::copy_if` algorithm are declared:

```
template <std::input_iterator I,
          std::sentinel_for<I> S,
          std::weakly_incrementable O,
          class Proj = std::identity,
          std::indirect_unary_predicate<
              std::projected<I, Proj>> Pred>
requires std::indirectly_copyable<I, O>
constexpr copy_if_result<I, O> copy_if(I first, S last,
                                       O result,
                                       Pred pred,
                                       Proj proj = {} );


template <ranges::input_range R,
```

```
        std::weakly_incrementable O,
        class Proj = std::identity,
        std::indirect_unary_predicate<
        std::projected<ranges::iterator_t<R>, Proj>> Pred>
  requires std::indirectly_copyable<ranges::iterator_t<R>, O>
  constexpr copy_if_result<ranges::borrowed_iterator_t<R>, O>
        copy_if(R&& r,
                O result,
                Pred pred,
                Proj proj = {});
```

If these seem more difficult to read, it is because they have more arguments, constraints, and longer type names. The good part, however, is that they make the code easier to write. Here is the previous snippet rewritten to use `std::ranges::copy_if`:

```
std::vector<int> v{ 1, 1, 2, 3, 5, 8, 13 };
std::vector<int> o;
auto e1 = std::ranges::copy_if(v, std:::back_inserter(o),
                               1_odd);

int arr[] = { 1, 1, 2, 3, 5, 8, 13 };
auto e2 = std::ranges::copy_if(arr, std:::back_inserter(o),
                               1_odd);

auto r = std::ranges::views::iota(1, 10);
auto e3 = std::ranges::copy_if(r, std:::back_inserter(o),
                               1_odd);
```

These examples show two things: how to copy elements from a `std::vector` object and an array and how to copy elements from a view (a range adaptor). What they don't show is projections. This was briefly mentioned earlier. We'll discuss it with more details and examples here.

A projection is an invocable entity. It's basically a function adaptor. It affects the predicate, providing a way to perform function composition. It does not provide a way to change the algorithm. For instance, let's say we have the following type:

```
struct Item
{
```

```
    int         id;
    std::string name;
    double      price;
};
```

Also, for the purpose of the explanation, let's also consider the following sequence of elements:

```
std::vector<Item> items{
    {1, "pen", 5.49},
    {2, "ruler", 3.99},
    {3, "pensil case", 12.50}
};
```

Projections allow you to perform composition on the predicate. For instance, let's say we want to copy to a second vector all the items whose names begin with the letter *p*. We can write the following:

```
std::vector<Item> copies;
std::ranges::copy_if(
    items,
    std::back_inserter(copies),
    [](Item const& i) {return i.name[0] == 'p'; });
```

However, we can also write the following equivalent example:

```
std::vector<Item> copies;
std::ranges::copy_if(
    items,
    std::back_inserter(copies),
    [](std::string const& name) {return name[0] == 'p'; },
    &Item::name);
```

The projection, in this example, is the pointer-to-member expression &Item::name that is applied to each Item element before executing the predicate (which is a lambda expression here). This can be useful when you already have reusable function objects or lambda expressions and you don't want to write another one for passing different types of arguments.

What projects cannot be used for, in this manner, is transforming a range from one type into another. For instance, you cannot just copy the names of the items from `std::vector<Item>` to `std::vector<std::string>`. This requires the use of the `std::ranges::transform` range adaptor, as shown in the following snippet:

```
std::vector<std::string> names;
std::ranges::copy_if(
    items | rv::transform(&Item::name),
    std::back_inserter(names),
    [](std::string const& name) {return name[0] == 'p'; });
```

There are many constrained algorithms, but we will not list them here. Instead, you can check them all either directly in the standard, or on the `https://en.cppreference.com/w/cpp/algorithm/ranges` page.

The last topic that we'll address in this chapter is writing a custom range adaptor.

Writing your own range adaptor

The standard library contains a series of range adaptors that can be used for solving many different tasks. More are being added in newer versions of the standard. However, there can be situations when you'd like to create your own range adaptor to use with others from the range library. This is not actually a trivial task. For this reason, in this final section of the chapter, we will explore the steps you need to follow to write such a range adaptor.

For this purpose, we will consider a range adaptor that takes every *Nth* element of a range and skips the others. We will call this adaptor `step_view`. We can use it to write code as follows:

```
for (auto i : std::views::iota(1, 10) | views::step(1))
    std::cout << i << '\n';

for (auto i : std::views::iota(1, 10) | views::step(2))
    std::cout << i << '\n';

for (auto i : std::views::iota(1, 10) | views::step(3))
    std::cout << i << '\n';

for (auto i : std::views::iota(1, 10) | views::step(2) |
```

```
                     std::views::take(3))
    std::cout << i << '\n';
```

The first loop will print all the numbers from one to nine. The second loop will print all the odd numbers, 1, 3, 5, 7, 9. The third loop will print 1, 4, 7. Lastly, the fourth loop will print 1, 3, 5.

To make this possible, we need to implement the following entities:

- A class template that defines the range adaptor

- A deduction guide to help with class template argument deduction for the range adaptor

- A class template that defines the iterator type for the range adaptor

- A class template that defines the sentinel type for the range adaptor

- An overloaded pipe operator (|) and helper functors, required for its implementation

- A compile-time constant global object to simplify the use of the range adaptor

Let's take them one by one and learn how to define them. We'll start with the sentinel class. A **sentinel** is an abstraction of a past-the-end iterator. It allows us to check whether an iteration reached the end of a range. A sentinel makes it possible for the end iterator to have a different type than the range iterators. Sentinels cannot be dereferenced or incremented. Here is how it can be defined:

```
template <typename R>
struct step_iterator;

template <typename R>
struct step_sentinel
{
    using base      = std::ranges::iterator_t<R>;
    using size_type = std::ranges::range_difference_t<R>;

    step_sentinel() = default;

    constexpr step_sentinel(base end) : end_{ end } {}
    constexpr bool is_at_end(step_iterator<R> it) const;
```

```
private:
    base        end_;
};


// definition of the step_iterator type


template <typename R>
constexpr bool step_sentinel<R>::is_at_end(
    step_iterator<R> it) const
{
    return end_ == it.value();
}
```

The sentinel is constructed from an iterator and contains a member function called
`is_at_end` that checks whether the stored range iterator is equal to the range iterator
stored in a `step_iterator` object. This type, `step_iterator`, is a class template
that defines the iterator type for our range adaptor, which we call `step_view`. Here is an
implementation of this iterator type:

```
template <typename R>
struct step_iterator : std::ranges::iterator_t<R>
{
    using base
        = std::ranges::iterator_t<R>;
    using value_type
        = typename std::ranges::range_value_t<R>;
    using reference_type
        = typename std::ranges::range_reference_t<R>;

    constexpr step_iterator(
        base start, base end,
        std::ranges::range_difference_t<R> step) :
        pos_{ start }, end_{ end }, step_{ step }
    {
    }

    constexpr step_iterator operator++(int)
```

```
    {
        auto ret = *this;
        pos_ = std::ranges::next(pos_, step_, end_);
        return ret;
    }

    constexpr step_iterator& operator++()
    {
        pos_ = std::ranges::next(pos_, step_, end_);
        return *this;
    }

    constexpr reference_type operator*() const
    {
        return *pos_;
    }

    constexpr bool operator==(step_sentinel<R> s) const
    {
        return s.is_at_end(*this);
    }

    constexpr base const value() const { return pos_; }

private:
    base                             pos_;
    base                             end_;
    std::ranges::range_difference_t<R>  step_;
};
```

This type must have several members:

- The alias template called base that represents the type of the underlying range iterator.

- The alias template called value_type that represents the type of elements of an underlying range.

- The overloaded operators ++ and *.
- The overloaded operator == compares this object with a sentinel.

The implementation of the ++ operator uses the std::ranges::next constrained algorithm to increment an iterator with *N* positions, but not past the end of the range.

In order to use the step_iterator and step_sentinel pair for the step_view range adaptor, you must make sure this pair is actually well-formed. For this, we must ensure that the step_iterator type is an input iterator, and that the step_sentinel type is indeed a sentinel type for the step_iterator type. This can be done with the help of the following static_assert statements:

```
namespace details
{
    using test_range_t =
        std::ranges::views::all_t<std::vector<int>>;
    static_assert(
        std::input_iterator<step_iterator<test_range_t>>);
    static_assert(
        std::sentinel_for<step_sentinel<test_range_t>,
        step_iterator<test_range_t>>);
}
```

The step_iterator type is used in the implementation of the step_view range adaptor. At a minimum, this could look as follows:

```
template<std::ranges::view R>
struct step_view :
    public std::ranges::view_interface<step_view<R>>
{
private:
    R                                  base_;
    std::ranges::range_difference_t<R>  step_;

public:
    step_view() = default;

    constexpr step_view(
        R base,
```

```cpp
         std::ranges::range_difference_t<R> step)
            : base_(std::move(base))
            , step_(step)
    {
    }


    constexpr R base() const&
        requires std::copy_constructible<R>
    { return base_; }
    constexpr R base()&& { return std::move(base_); }


    constexpr std::ranges::range_difference_t<R> const&
increment() const
    { return step_; }

    constexpr auto begin()
    {
        return step_iterator<R const>(
            std::ranges::begin(base_),
            std::ranges::end(base_), step_);
    }


    constexpr auto begin() const
    requires std::ranges::range<R const>
    {
        return step_iterator<R const>(
            std::ranges::begin(base_),
            std::ranges::end(base_), step_);
    }

    constexpr auto end()
    {
        return step_sentinel<R const>{
            std::ranges::end(base_) };
    }
```

```
    constexpr auto end() const
    requires std::ranges::range<R const>
    {
        return step_sentinel<R const>{
            std::ranges::end(base_) };
    }

    constexpr auto size() const
    requires std::ranges::sized_range<R const>
    {
        auto d = std::ranges::size(base_);
        return step_ == 1 ? d :
            static_cast<int>((d + 1)/step_); }

    constexpr auto size()
    requires std::ranges::sized_range<R>
    {
        auto d = std::ranges::size(base_);
        return step_ == 1 ? d :
            static_cast<int>((d + 1)/step_);
    }
};
```

There is a pattern that must be followed when defining a range adaptor. This pattern is represented by the following aspects:

- The class template must have a template argument that meets the `std::ranges::view` concept.

- The class template should be derived from `std::ranges:view_interface`. This takes a template argument itself and that should be the range adaptor class. This is basically an implementation of the CRTP that we learned about in *Chapter 7, Patterns and Idioms*.

- The class must have a default constructor.

- The class must have a `base` member function that returns the underlying range.

- The class must have a `begin` member function that returns an iterator to the first element in the range.

- The class must have an end member function that returns either an iterator to the one-past-the-last element of the range or a sentinel.

- For ranges that meet the requirements of the `std::ranges::sized_range` concept, this class must also contain a member function called `size` that returns the number of elements in the range.

In order to make it possible to use class template argument deduction for the `step_view` class, a user-defined deduction guide should be defined. These were discussed in *Chapter 4, Advanced Template Concepts*. Such a guide should look as follows:

```
template<class R>
step_view(R&& base,
          std::ranges::range_difference_t<R> step)
   -> step_view<std::ranges::views::all_t<R>>;
```

In order to make it possible to compose this range adaptor with others using the pipe iterator (|), this operator must be overloaded. However, we need some helper function object, which is shown in the next listing:

```
namespace details
{
    struct step_view_fn_closure
    {
        std::size_t step_;
        constexpr step_view_fn_closure(std::size_t step)
            : step_(step)
        {
        }

        template <std::ranges::range R>
        constexpr auto operator()(R&& r) const
        {
            return step_view(std::forward<R>(r), step_);
        }
    };

    template <std::ranges::range R>
    constexpr auto operator | (R&& r,
```

```
                                   step_view_fn_closure&& a)
   {
      return std::forward<step_view_fn_closure>(a)(
         std::forward<R>(r));
   }
}
```

The `step_view_fn_closure` class is a function object that stores a value representing the number of elements to skip for each iterator. Its overloaded call operator takes a range as an argument and returns a `step_view` object created from the range and the value for the number of steps to jump.

Finally, we want to make it possible to write code in a similar manner to what is available in the standard library, which provides a compile-time global object in the `std::views` namespace for each range adaptor that exists. For instance, instead of `std::ranges::transform_view`, you could use `std::views::transform`. Similarly, instead of `step_view` (in some namespace), we want to have an object, `views::step`. To do so, we need yet another function object, as shown next:

```
namespace details
{
   struct step_view_fn
   {
      template<std::ranges::range R>
      constexpr auto operator () (R&& r,
                                    std::size_t step) const
      {
         return step_view(std:::forward<R>(r), step);
      }

      constexpr auto operator () (std::size_t step) const
      {
         return step_view_fn_closure(step);
      }
   };
}

namespace views
{
```

```
    inline constexpr details::step_view_fn step;
}
```

The `step_view_fn` type is a function object that has two overloads for the call operator: one takes a range and an integer and returns a `step_view` object, and the other takes an integer and returns a closure for this value, or, more precisely, an instance of `step_view_fn_closure` that we saw earlier.

Having all these implemented, we can successfully run the code shown at the beginning of this section. We have completed the implementation of a simple range adaptor. Hopefully, this should give you a sense of what writing range adaptors takes. The ranges library is significantly complex when you look at the details. In this chapter, you have learned some basics about the content of the library, how it can simplify your code, and how you can extend it with custom features. This knowledge should be a starting point for you should you want to learn more using other resources.

Summary

In this final chapter of the book, we explored the C++20 ranges library. We started the discussion with a transition from the abstract concept of a range to the new ranges library. We learned about the content of this library and how it can help us write simpler code. We focused the discussion on range adapters but also looked at constrained algorithms. At the end of the chapter, we learned how to write a custom range adaptor that can be used in combination with standard adapters.

Questions

1. What is a range?
2. What is a view in the range library?
3. What are constrained algorithms?
4. What is a sentinel?
5. How can you check that a sentinel type corresponds to an iterator type?

Further reading

- *A beginner's guide to C++ Ranges and Views*, Hannes Hauswedell, `https://hannes.hauswedell.net/post/2019/11/30/range_intro/`

- *Tutorial: Writing your first view from scratch (C++20/P0789)*, Hannes Hauswedell, `https://hannes.hauswedell.net/post/2018/04/11/view1/`

- *C++20 Range Adaptors and Range Factories*, Barry Revzin, `https://brevzin.github.io/c++/2021/02/28/ranges-reference/`

- *Implementing a better views::split*, Barry Revzin, `https://brevzin.github.io/c++/2020/07/06/split-view/`

- *Projections are Function Adaptors*, Barry Revzin, `https://brevzin.github.io/c++/2022/02/13/projections-function-adaptors/`

- *Tutorial: C++20's Iterator Sentinels*, Jonathan Müller, `https://www.foonathan.net/2020/03/iterator-sentinel/`

- *Standard Ranges*, Eric Niebrel, `https://ericniebler.com/2018/12/05/standard-ranges/`

- *Zip*, Tim Song, `http://www.open-std.org/jtc1/sc22/wg21/docs/papers/2021/p2321r2.html`

- *From range projections to projected ranges*, Oleksandr Koval, `https://oleksandrkvl.github.io/2021/10/11/projected-ranges.html`

- *Item 30 - Create custom composable views*, Wesley Shillingford, `https://cppuniverse.com/EverydayCpp20/RangesCustomViews`

- *A custom C++20 range view*, Marius Bancila, `https://mariusbancila.ro/blog/2020/06/06/a-custom-cpp20-range-view/`

- *New C++23 Range Adaptors*, Marius Bancila, `https://mariusbancila.ro/blog/2022/03/16/new-cpp23-range-adaptors/`

- *C++ Code Samples Before and After Ranges*, Marius Bancila, `https://mariusbancila.ro/blog/2019/01/20/cpp-code-samples-before-and-after-ranges/`

Appendix
Closing Notes

We are now at the end of this book. Templates are not the easiest part of C++ programming. Indeed, people usually find them difficult or horrendous. However, templates are heavily used in C++ code, and it's likely that whatever kind of code you're writing, you'll be using templates daily.

We started the book by learning what templates are and why we need them. We then learned how to define function templates, class templates, variable templates, and alias templates. We learned about template parameters, specialization, and instantiation. In the third chapter, we learned about templates with variable numbers of arguments, which are called variadic templates. The next chapter was dedicated to more advanced template concepts, such as name binding, recursion, argument deduction, and forwarding references.

We then learned about the use of type traits, SFINAE, and constexpr if, and we explored the collection of type traits available in the standard library. The sixth chapter was dedicated to concepts and constraints, which are part of the C++20 standard. We learned how to specify constraints for template arguments in different ways and how to define concepts, and everything related to them. We also explored the collection of concepts available in the standard library.

In the final part of the book, we focused on using templates for practical purposes. First, we explored a series of patterns and idioms, such as the CRTP, mixins, type erasure, tag dispatching, expression templates, and type lists. Then, we learned about containers, iterators, and algorithms, which are the pillars of the Standard Template Library, and wrote some of our own. Finally, the last chapter was dedicated to the C++20 ranges library, where we learned about ranges, range adaptors, and constrained algorithms.

By reaching this point, you have completed this journey of learning metaprogramming with C++ templates. However, this learning process doesn't end here. A book can only supply you with the necessary information to learn a topic, structured in such a way that makes it easy to understand and follow that topic. But reading a book without practicing what you have learned is futile. Your task now is to put into practice the knowledge you have acquired from this book at work, at school, or at home. Because only by practicing will you be able to truly master not only the C++ language and metaprogramming with templates but also any other skill.

I hope this book will prove to be a valuable resource for you in reaching the goal of becoming prolific with C++ templates. While developing this book, I tried to find the right balance between simplicity and meaningfulness so that it makes it easier for you to learn this difficult topic. I hope I succeeded in doing that.

Thank you for reading this book, and I wish you good luck putting it into practice.

Assignment Answers

Chapter 1, Introduction to Templates

Question 1

Why do we need templates? What advantages do they provide?

Answer

There are several benefits to using templates: they help us avoid writing repetitive code, they foster the creation of generic libraries, and they can help us write less and better code.

Question 2

How do you call a function that is a template? What about a class that is a template?

Answer

A function that is a template is called a function template. Similarly, a class that is a template is called a class template.

Question 3

How many kinds of template parameters exist and what are they?

Answer

There are three kinds of template parameters: type template parameters, non-type template parameters, and template template parameters.

Question 4

What is partial specialization? What about full specialization?

Answer

Specialization is the technique of providing an alternative implementation for a template, called the primary template. Partial specialization is an alternative implementation provided for only some of the template parameters. A full specialization is an alternative implementation when arguments are provided for all the template parameters.

Question 5

What are the main disadvantages of using templates?

Answer

The main disadvantages of using templates include the following: complex and cumbersome syntax, compiler errors that are often long and hard to read and understand, and increased compilation times.

Chapter 2, Template Fundamentals

Question 1

What category of types can be used for non-type template parameters?

Answer

Non-type template parameters can only have structural types. Structure types are integral types, floating-point types (as of C++20), enumeration types, pointer types (either to objects or functions), pointer to member types (either to member objects or member functions), lvalue reference types (either to objects or functions), and literal class types that meet several requirements: all base classes are public and non-mutable, all non-static data members are public and non-mutable, and the types of all the base classes and the non-static data members are also structural types or arrays thereof. `const` and `volatile` qualified versions of these types are also allowed.

Question 2

Where are default template arguments not allowed?

Answer

Default template arguments cannot be used for parameter packs, in declarations of friend class templates, and in the declaration or definition of an explicit specialization of a function template or member function template.

Question 3

What is explicit instantiation declaration and how does it differ syntactically from explicit instantiation definition?

Answer

Explicit instantiation declaration is the way you can tell the compiler that the definition of a template instantiation is found in a different translation unit and that a new definition should not be generated. The syntax is the same as for explicit instantiation definitions, except that the `extern` keyword is used in front of the declaration.

Question 4

What is an alias template?

Answer

An alias template is a name that, unlike type aliases, which refer to another type, refers to a template or, in other words, a family of types. Alias templates are introduced with using declarations. They cannot be introduced with `typedef` declarations.

Question 5

What are template lambdas?

Answer

Template lambdas are an improved form of generic lambdas, introduced in C++20. They allow us to use the template syntax to explicitly specify the shape of the templatized function-call operator of the function object that the compiler is generating for a lambda expression.

Chapter 3, Variadic Templates

Question 1

What are variadic templates and why are they useful?

Answer

Variadic templates are templates with a variable number of arguments. They allow us to write not only functions with variable number of arguments but also class templates, variable templates, and alias templates. Unlike other approaches, such as the use of the `va_` macros, they are type-safe, do not require macros, and do not require us to explicitly specify the number of arguments.

Question 2

What is a parameter pack?

Answer

There are two kinds of parameter packs: template parameter packs and function parameter packs. The former are template parameters that accept zero, one, or more template arguments. The latter are function parameters that accept zero, one, or more function arguments.

Question 3

What are the contexts where parameter packs can be expanded?

Answer

Parameter packs can be expanded in a multitude of contexts, as follows: template parameter lists, template argument lists, function parameter lists, function argument lists, parenthesized initializers, brace-enclosed initializers, base specifiers and member initializer lists, fold expressions, using declarations, lambda captures, the `sizeof...` operator, alignment specifiers, and attribute lists.

Question 4

What are fold expressions?

Answer

A fold expression is an expression involving a parameter pack that folds (or reduces) the elements of the parameter pack over a binary operator.

Question 5

What are the benefits of using fold expressions?

Answer

The benefits of using fold expressions include having less and simpler code to write, fewer template instantiations, which lead to faster compile times, and potentially faster code, since multiple function calls are replaced with a single expression.

Chapter 4, Advanced Template Concepts

Question 1

When is name lookup performed?

Answer

Name lookup is performed at the point of template instantiation for dependent names (those that depend on the type or value of a template parameter) and at the point of template definition for non-dependent names (those that don't depend on template parameters).

Question 2

What are deduction guides?

Answer

Deduction guides are a mechanism that tells the compiler how to perform class template argument deduction. Deduction guides are fictional function templates representing constructor signatures of a fictional class type. If overload resolution fails on the constructed set of fictional function templates, then the program is ill-formed and an error is generated. Otherwise, the return type of the selected function template specialization becomes the deduced class template specialization.

Question 3

What are forwarding references?

Answer

A forward reference (also known as universal reference) is a reference in a template that behaves as an rvalue reference if an rvalue was passed as an argument or an lvalue reference if an lvalue was passed as an argument. A forwarding reference must have the `T&&` form such as in `template <typename T> void f(T&&)`. Forms such as `T const &&` or `std::vector<T>&&` do not represent forwarding references but normal rvalue references.

Question 4

What does `decltype` do?

Answer

The `decltype` specifier is a type specifier. It returns the type of an expression. It is usually used in templates together with the `auto` specifier in order to declare the return type of a function template that depends on its template arguments, or the return type of a function that wraps another function and returns the result from executing the wrapped function.

Question 5

What does `std::declval` do?

Answer

`std::declval` is a utility function template from the `<utility>` header that adds an rvalue reference to its type template argument. It can only be used in unevaluated contexts (compile-time-only contexts that are not evaluated during runtime), and its purpose is to help with dependent type evaluation for types that do not have a default constructor or one that cannot be accessed because it's private or protected.

Chapter 5, Type Traits and Conditional Compilation

Question 1

What are type traits?

Answer

Type traits are small class templates that enable us to either query properties of types or perform transformations of types.

Question 2

What is SFINAE?

Answer

SFINAE is an acronym for **Substitution Failure Is Not An Error**. This is a rule for template substitution and works as follows: when the compiler encounters the use of a function template, it substitutes the arguments in order to instantiate the template; if an error occurs at this point, it is not regarded as an ill-formed code, only as a deduction failure. As a result, the function is removed from the overload set instead of causing an error. Therefore, an error only occurs if there is no match in the overload set for a particular function call.

Question 3

What is `constexpr if`?

Answer

`constexpr if` is a compile-time version of the `if` statement. The syntax for it is `if constexpr(condition)`. It's been available since C++17 and allows us to discard a branch, at compile time, based on the value of a compile-time expression.

Question 4

What does `std::is_same` do?

Answer

`std::is_same` is a type trait that checks whether two types are the same. It includes checks for the `const` and `volatile` qualifiers, yielding `false` for two types that have different qualifiers (such as `int` and `int const`).

Question 5

What does `std::conditional` do?

Answer

`std::conditional` is a metafunction that chooses one type or another based on a compile-time constant.

Chapter 6, Concepts and Constraints

Question 1

What are constraints? What about concepts?

Answer

A constraint is a requirement imposed on a template argument. A concept is a named set of one or more constraints.

Question 2

What is a requires clause and a requires expression?

Answer

A requires clause is a construct that allows us to specify a constraint on a template argument or function declaration. This construct is composed of the `requires` keyword followed by a compile-time Boolean expression. A requires clause affects the behavior of a function, including it for overload resolution only if the Boolean expression is `true`. On the other hand, a requires expression has the `requires (parameters-list) expression;` form, where `parameters-list` is optional. Its purpose is to verify that some expressions are well-formed, without having any side effects or affecting the behavior of the function. Requires expressions can be used with requires clauses, although named concepts are preferred, mainly for readability.

Question 3

What are the categories of requires expressions?

Answer

There are four categories of requires expressions: simple requirements, type requirements, compound requirements, and nested requirements.

Question 4

How do constraints affect the ordering of templates in overload resolution?

Answer

The constraining of functions affects their order in the overload resolution set. When multiple overloads match the set of arguments, the overload that is more constrained is selected. However, keep in mind that constraining with type traits (or Boolean expressions in general) and concepts is not semantically equal. For details on this topic, revisit the *Learning about the ordering of templates with constraints* section.

Question 5

What are abbreviated function templates?

Answer

Abbreviated function templates are a new feature introduced in C++20 that provides a simplified syntax for function templates. The `auto` specifier can be used to define function parameters and the template syntax can be skipped. The compiler will automatically generate a function template from an abbreviated function template. Such functions can be constrained using concepts, therefore imposing requirements on the template arguments.

Chapter 7, Patterns and Idioms

Question 1

What are typical problems for which the Curiously Recuring Template Pattern is used?

Answer

The **Curiously Recurring Template Pattern (CRTP)** is typically used for solving problems such as adding common functionality to types and avoiding code duplication, limiting the number of times a type can be instantiated, or implementing the composite design pattern.

Question 2

What are mixins and what is their purpose?

Answer

Mixins are small classes that are designed to add functionality to other classes, by inheriting from the classes they are supposed to complement. This is the opposite of the CRTP pattern.

Question 3

What is type erasure?

Answer

Type erasure is the term used to describe a pattern that removes information from types, making it possible for types that are not related to be treated in a generic way. Although forms of type erasure can be achieved with `void` pointers or polymorphism, the true type erasure pattern is achieved in C++ with templates.

Question 4

What is tag dispatching and what are its alternatives?

Answer

Tag dispatching is a technique that enables us to select one or another function overload at compile time. Although tag dispatching itself is an alternative to `std::enable_if` and SFINAE, it also has its own alternatives. These are constexpr if in C++17 and concepts in C++20.

Question 5

What are expression templates and where are they used?

Answer

Expression templates are a metaprogramming technique that enables a lazy evaluation of a computation at compile-time. The benefit of this technique is that it avoids performing inefficient operations at runtime at the expense of more complex code that could be difficult to comprehend. Expression templates are typically used to implement linear algebra libraries.

Chapter 8, Ranges and Algorithms

Question 1

What are the sequence containers from the standard library?

Answer

The sequence containers from the C++ standard library are `std::vector`, `std::deque`, `std::list`, `std::array`, and `std::forward_list`.

Question 2

What are the common member functions defined in the standard containers?

Answer

The member functions that are defined for most containers in the standard library are `size` (not present in `std::forward_list`), `empty`, `clear` (not present in `std::array`, `std::stack`, `std::queue`, and `std::priority_queue`), `swap`, `begin`, and `end`.

Question 3

What are iterators and how many categories exist?

Answer

An iterator is an abstraction that enables us to access the elements of a container in a generic way, without having to know the implementation details of each container. Iterators are key for writing general-purpose algorithms. There are six categories of iterators in C++: input, forward, bidirectional, random-access, contiguous (as of C++17), and output.

Question 4

What operations does a random-access iterator support?

Answer

The random-access iterators must support the following operations (in addition to those required for input, forward, and bidirectional iterators): the + and - arithmetic operators, inequality comparison (with other iterators), compound assignment, and offset dereference operators.

Question 5

What are range access functions?

Answer

Range access functions are non-member functions that provide a uniform way to access the data or properties of containers, arrays, and the `std::initializer_list` class. These functions include `std::size`/`std::ssize`, `std::empty`, `std::data`, `std::begin`, and `std::end`.

Chapter 9, The Ranges Library

Question 1

What is a range?

Answer

A range is an abstraction for a sequence of elements, defined with a beginning and end iterator. The beginning iterator points to the first element in the sequence. The end iterator points to the one-past-last element of the sequence.

Question 2

What is a view in the ranges library?

Answer

A view in the C++ ranges library, also called a range adaptor, is an object that implements an algorithm that takes one or more ranges as input and perhaps other arguments and returns an adapted range. Views are lazy-evaluated, meaning they do not perform the adaptation until their elements are iterated.

Question 3

What are constrained algorithms?

Answer

Constrained algorithms are implementations of the existing standard library algorithms but in the C++20 ranges library. They are called constrained because their template arguments are constrained using C++20 concepts. In these algorithms, instead of requiring a begin-end pair of iterators for specifying, a range of values accepts a single range argument. However, overloads that accept an iterator-sentinel pair also exist.

Question 4

What is a sentinel?

Answer

A sentinel is an abstraction for an end iterator. This makes it possible for the end iterator to have a different type than the range iterator. Sentinels cannot be dereferenced or incremented. Sentinels are useful when the test for the end of a range depends on some variable (dynamic) condition and you don't know you are at the end of the range until something happens (for instance, a condition becomes false).

Question 5

How can you check that a sentinel type corresponds to an iterator type?

Answer

You can check that a sentinel type can be used with an iterator type by using the `std::sentinel_for` concept from the `<iterator>` header.

Index

Symbols

|| operator 258
&& operator 258

A

abbreviated function template 276
abstraction 290
abstract ranges
  advancing, to ranges library 402-404
algorithm 370
alias template
  about 13, 68
  constraining 273, 274
  defining 67-70
allocator 368
Argument-Dependent Lookup (ADL)
  about 232
  reference link 232
atomic constraint 262
auto parameters
  constraining, with concepts 276-278

B

base 426
bidirectional iterator 372

C

C++ 290
C++17 420
C++20 130
C++23 407
c++ Insights
  reference link 161
C++ reference documentation
  reference link 284
callbacks 5
circular buffer 376
circular buffer container
  implementing 378-386
  iterator type, implementing 387-395
Clang 13 141
class template argument deduction
    12, 24, 155-162

class templates
  about 12
  constraining 271-273
  defining 23-25
client-attorney pattern 188
Compiler Explorer
  URL 86
compound requirements 252-256
concepts
  about 29, 245, 325
  defining 244-247
  need for 238-244
  used, for constraining auto
    parameters 276-278
  using 334, 335
conditionally terminated sequence 405
conjunction 258-261
connection class template 188
constant iterators 372
constexpr if
  using 209-213, 332, 333
constrained 278
constrained algorithms
  about 404, 419-423
  reference link 423
constraints
  about 29, 244
  composing 258-262
  ordering of templates 262-267
  specifying, ways 275
constraints normalization 263
containers
  about 366-369, 372-375
  categories 366
  creating 376, 377
copy algorithm
  implementing 229-232

counted sequence 404
Curiously Recurring Template
    Pattern (CRTP)
  about 294-296
  composite design pattern,
    implementing 301-307
  functionality, adding with 298-301
  in standard library 307-312
  object count, limiting with 296-298
current instantiation
  about 133-136
  names 134
custom range adaptor
  pattern 429
  writing 423-432
C with Classes 14

D

decltype specifier 170-178
default template arguments 40-43
dependent class 135
dependent names 122
dependent template names 131-133
dependent type 110
dependent type names 128-131
disjunction 258, 261
duck typing 321
dynamic polymorphism
  about 290
  versus static polymorphism 290-294

E

enable_if metafunction
  about 204
  usage, scenarios 205

enable_if type trait
   used, for enabling SFINAE 203-209
encapsulation 290
Excalibur 296
explicit instantiation
   about 47
   declaration 47, 51-53
   definition 47-50
explicit specialization 53-58
expression templates
   about 336-345
   ranges library, using as
      alternative 345, 346

F

factories 408
fold expression
   about 110-113
   benefits 114
forwarding references 162-170
forward iterator 390
friends 181
functionality
   adding, with CRTP 298-301
function parameter pack 85
function template 12
function template argument
      deduction 142-154
function templates
   defining 20-22

G

game unit 349
GCC 12 141
general-purpose algorithm
   writing 395-398

generic lambda
   about 71, 276
   exploring 70-78

H

homogenous variadic function template
   building 233-235

I

immediate context 199
implicit instantiation 43-47
inheritance 290
iterators
   about 369-375
   categories 370
   creating 376, 377
iterator type
   implementing, for circular
      buffer container 387-395

J

Java 318

L

lambda expressions 70
lambda templates
   about 71
   exploring 70-78
lvalue references 163
lvalues 162

M

member function templates
    defining 26-28
metafunctions 203, 223
mixins 312-318
model 325
modern C++ 15
move semantics 162
mutable iterator 372

N

name binding 122-124
nested requirements 256, 257
non-dependent class 135
non-dependent names 122
non-template member functions
    constraining 267-271
non-type template parameters 30-37

O

object count
    limiting, with CRTP 296-298
operations
    implementing, on typelists 352-361
ordering, of templates
    with constraints 262-267
output iterator 390

P

parameter pack
    about 29, 90-94
    alignment specifier 100
    attribute list 100
    expansion 95

parameter pack expansion
    about 85
    base specifiers and member
        initializer lists 98
    brace-enclosed initializers 97
    declarations, using 98, 99
    fold-expressions 99
    function argument list 96
    parenthesized initializers 96
    template argument list 95
    template parameter list 95
partial specialization 58-62
polymorphic classes 290, 291
polymorphism 290
primary template 53
projection 420
Python 318

R

random-access iterator 372, 391
range
    about 345, 374, 402, 404
    concepts 405, 406
    sequence, defining 404
    view 407-414
range access functions 406
ranges library
    abstract ranges, advancing to 402-404
    constrained algorithms 404, 419-423
    factories 408
    using, as alternative to expression
        templates 345, 346
    views 404
range-v3 library
    reference link 345
reference collapsing 168
requires clause 242, 245, 275

requires expressions
  about 245, 275
  compound requirements 252-256
  exploring 248
  nested requirements 256, 257
  simple requirement 248-250
  type requirements 251, 252
  types 248
rvalue references 162, 163
rvalues 162

S

sentinel 404, 424
serializer class template 197
simple requirement 248-250
size, member function 430
standard concepts library
  exploring 279-284
standard library
  CRTP 307-312
  examples 327, 328
Standard Template Library (STL) 15
standard type traits
  cv-specifiers, modifying 223, 224
  exploring 213
  miscellaneous transformations 224-228
  pointers, modifying 223, 224
  references, modifying 223, 224
  sign, modifying 223, 224
  supported operations, querying 220, 221
  type category, querying 214-217
  type properties, querying 217-220
  type relationships, querying 222, 223
static polymorphism
  about 290, 292
  versus dynamic polymorphism 290-293

std::declval type
  operator 179, 180
step_view_fn_closure 431
step_view_fn type 432
structural types 30
Substitution Failure Is Not An
    Error (SFINAE)
  about 197, 328
  enabling, with enable_if
      type trait 203-209
  exploring 197-202
  purpose 198-202
sum 238

T

tag dispatching 328-331
tag dispatching, alternatives
  about 332
  concepts, using 334, 335
  constexpr if, using 332, 333
take_view 409
template argument deduction
  for class 155-162
  for functions 142-154
template instantiation
  about 13, 43
  explicit instantiation 47
  forms 14
  implicit instantiation 43-47
template parameter pack 85
template parameters
  about 13, 28
  categories 13
  default template arguments 40-43
  non-type template parameters 30-37
  type template parameters 28, 29

template recursion
  exploring 136-142
templates
  cons 16, 17
  friend operators 181-188
  history 14-16
  need for 4-8
  pros 16, 17
  terminology 12-14
  writing 8-12
template specialization
  about 53
  explicit specialization 53-58
  partial specialization 58-62
template template parameters 38, 39
two-phase name lookup 125-127
type category
  querying 214-217
type erasure 318-328
typelists
  about 347-349
  operations, implementing 352-361
  using 349-351
type properties
  querying 217-220
type relationships
  querying 222, 223
type requirements
  about 251, 252
  usage 251
type template parameter 8, 28, 29
type template parameter pack 29
type traits
  about 192-197
  categories 203
  using, examples 228

type traits, using examples
  copy algorithm, implementing 229-232
  homogenous variadic function
      template, building 233-235

U

unbounded sequence 405
unconstrained 278
unevaluated context 173
universal references 166, 170
user-defined deduction guides 158

V

value categories
  reference link 162
value_type 426
variable template
  about 13
  constraining 273, 274
  defining 63-67
variadic alias templates
  about 114, 115
  index_sequence 116
  index_sequence_for 116
  make_index_sequence 116
variadic class templates 101-110
variadic function templates
  about 84-88
  implementing 85-89
variadic templates
  code writing, drawbacks 83
  need for 82-84
variadic variable templates 116
VC++ 16.11 141

vector class
  implementation 337
views
  about 404, 407-410
  examples, exploring 414-418
  listing 410-414
void pointers 318

W

wrapper class 181-187

Z

zipping algorithm 395

Other Books You May Enjoy

If you enjoyed this book, you may be interested in these other books by Packt:

C++20 STL Cookbook

Bill Weinman

ISBN: 978-1-80324-871-4

- Understand the new language features and the problems they can solve
- Implement generic features of the STL with practical examples
- Understand standard support classes for concurrency and synchronization
- Perform efficient memory management using the STL
- Implement seamless formatting using std::format
- Work with strings the STL way instead of handcrafting C-style code

CMake Best Practices

Dominik Berner, Mustafa Kemal Gilor

ISBN: 978-1-80323-972-9

- Get to grips with architecting a well-structured CMake project
- Modularize and reuse CMake code across projects
- Integrate various tools for static analysis, linting, formatting, and documentation into a CMake project
- Get hands-on with performing cross-platform builds
- Discover how you can easily use different toolchains with CMake
- Get started with crafting a well-defined and portable build environment for your project

Packt is searching for authors like you

If you're interested in becoming an author for Packt, please visit authors. packtpub.com and apply today. We have worked with thousands of developers and tech professionals, just like you, to help them share their insight with the global tech community. You can make a general application, apply for a specific hot topic that we are recruiting an author for, or submit your own idea.

Share Your Thoughts

Now you've finished *Template Metaprogramming with C++*, we'd love to hear your thoughts! Scan the QR code below to go straight to the Amazon review page for this book and share your feedback or leave a review on the site that you purchased it from.

https://packt.link/r/1803243457

Your review is important to us and the tech community and will help us make sure we're delivering excellent quality content.

www.ingramcontent.com/pod-product-compliance
Lightning Source LLC
Chambersburg PA
CBHW081456050326
40690CB00015B/2814